The Life Cycle

The Life Cycle
Readings in Human Development

Laurence D. Steinberg
with the assistance of Lynn J. Mandelbaum

Columbia University Press   New York

For permission to reprint copyrighted material, acknowledgment is made to the publishers, organizations, and authors named on pp. xi-xiv, which constitute an extension of this copyright page.

Library of Congress Cataloging in Publication Data
Main entry under title:

The Life Cycle.

  Includes bibliographies.
  1. Life cycle, Human—Addresses, essays, lectures.
2. Life cycle, Human—Psychological aspects—Addresses,
essays, lectures. 3. Identity (Psychology)—Addresses,
essays, lectures. 4. Interpersonal relations—Addresses,
essays, lectures. 5. Performance—Addresses, essays,
lectures. I. Steinberg, Laurence D., 1952–
II. Mandelbaum, Lynn J.
HQ799.5.L54   305   81-3806
ISBN 0-231-05110-7   AACR2
ISBN 0-231-05111-5 (pbk.)

Columbia University Press
New York   Guilford, Surrey

Contents

Part IV: Adulthood

MASTERY AND COMPETENCIES

IDENTITY AND THE SELF

RELATIONS WITH OTHERS

Part V: Late Adulthood

MASTERY AND COMPETENCIES

IDENTITY AND THE SELF

RELATIONS WITH OTHERS

Preface

THE PURPOSE OF THIS BOOK is to provide the student of human development with a collection of empirical and theoretical articles reflecting both contemporary and classic thinking in the field. Although a good textbook can offer a thorough summary of information on development and, in some cases, a sensible organizational framework within which to place this information, there is really no substitute for reading important contributions in the original. Above all, human development is a science, and it is important for the student to see how theories are developed and how our ideas about development are tested empirically.

The study of individual development over the life span has grown dramatically within the past decade, and it is now common to find that training in life-span human development is required in programs in psychology, social work, nursing, counseling, education, and medicine. The field of life-span development has three concerns, all of which are reflected in this anthology: (1) how individuals grow and change over the course of the life-cycle—influences on development; (2) distinguishing characteristics of different periods in the life cycle—predictable tasks, events, and developments that differentiate one period from the next; and (3) themes and issues that recur throughout the entire life span—developmental issues that may change form over the life cycle but never lose significance.

It is, of course, impossible to cover every aspect of development in a collection of articles that a student might reasonably be expected to read during a course on human development. I have tried to include articles that deal with what I consider to be the central developmental issues of each period in the life cycle in a way that reflects development in the real world. Although the vast majority of the selections were written within the past decade, I did not hesitate to include older selections that are still influential. Few thinkers, for example, have had the impact of Erik Erikson or J. McVicker Hunt, and their ideas are as important today as they were twenty years ago.

The organization of the book illustrates my own belief that, although it is possible to separate the life cycle into discrete periods such as infancy or adolescence, the fundamental developmental issues are continuous: developing and maintaining mastery and competence in dealing with the environment; expanding and reorganizing one's sense of self and identity; and creating and responding to changes in relationships with significant others. Thus, the infant's exploration of the environment during the first two years of life can be viewed in some respects as we might view a middle-aged adult's response to work; in both cases, we might ask whether and in what ways the environment contributes to, or interferes with, the individual's developing sense of mastery.

The collection is organized around five periods in the life cycle—infancy, childhood, adolescence, adulthood, and late adulthood—and each section includes six selections drawn from three developmental domains: *Mastery and Competencies, Identity and the Self,* and *Relations with Others.* I have tried to create an organization that can be easily adapted to either an issue-oriented course (in which particular aspects of development, such as identity development, are traced across the entire life span one by one) or a period-oriented course (in which each period in the life cycle is discussed in its entirety one by one).

I was assisted in compiling this anthology by several colleagues, to whom I am indebted. Lynn J. Mandelbaum, whose contribution is acknowledged on the title page, was an invaluable consultant during the process of narrowing down the field of hundreds of potential inclusions to the thirty articles ultimately selected. Jay Belsky, Jodie Korbin, and Deborah Stipek all made helpful suggestions about specific articles that I might have otherwise overlooked. Joan Gordon handled much of the correspondence and clerical work that a project of this nature requires. Finally, I wish to thank the University of California's Committee on Instructional Development, whose funding supported much of this project during its early stages.

Laurence D. Steinberg
Irvine, California
May 1980

Acknowledgments

I WOULD LIKE TO THANK THE following for permission to reprint materials used in this book. It should be noted that, to achieve uniformity, in some cases original reference and/or footnote formats were altered to conform with the author-date method of citation used throughout this book. It was also necessary to make some minor editorial changes appropriate to an anthology of this kind.

1. "Competent Newborns" by Tom Bower. In Roger Lewin, ed., *Child Alive!* Copyright © 1974, 1975 by IPC Magazines, Ltd. Used by permission of Doubleday & Company, Inc.
2. "Piaget's Observations as a Source of Hypotheses Concerning Motivation" by J. McVicker Hunt. From the *Merrill-Palmer Quarterly*, 1963, 9:263–275. Copyright © 1963 by the Merrill-Palmer Institute. Used by permission of the Merrill-Palmer Institute and the author.
3. "Mirror-Image Stimulation and Self-Recognition in Infancy" by Jeanne Brooks-Gunn and Michael Lewis. Based on a paper presented at the Society for Research in Child Development meetings, Denver, 1975. (Major portions are to be found in M. Lewis and J. Brooks-Gunn, *Social Cognition and the Acquisition of the Self*. New York: Plenum Press, 1979.) Copyright © 1979 by Jeanne Brooks-Gunn and Michael Lewis. Used by permission of the authors.
4. "Characteristics of the Individual Child's Behavioral Responses to the Environment" by Stella Chess, Alexander Thomas, and Herbert Birch. From *American Journal of Orthopsychiatry*, 1959, 29:791–802. Copyright © 1959 by the American Orthopsychiatric Association, Inc. Reproduced by permission.
5. "Attachment, Exploration, and Separation: Illustrated by the Behavior of One-Year Olds in a Strange Situation" by Mary D. Salter Ainsworth and Silvia M. Bell. From *Child Development*, 1970, 41:49–67. Copyright © 1970 by the Society for Research in Child Development. Used by permission of the Society for Research in Child Development and the authors.
6. "Growth of Social Play with Peers during the Second Year of Life" by

Carol O. Eckerman, Judith L. Whatley, and Stuart L. Kutz. From *Developmental Psychology*, 1975, 11(1):42–49. Copyright © 1975 by the American Psychological Association. Used by permission of the American Psychological Association and the authors.

7. "The Role of Play in Cognitive Development" by Brian Sutton-Smith. From *Young Children*, 1967, 22(6):361–370. Copyright © 1976 by the National Association for the Education of Young Children, 1834 Connecticut Avenue, N.W., Washington, D.C. 20009. Used by permission of the National Association for the Education of Young Children.

8. "A Structural-Developmental Analysis of Levels of Role-Taking in Middle Childhood" by Robert L. Selman and Diane F. Byrne. From *Child Development*, 1974, 45:803–806. Copyright © 1974 by the Society for Research in Child Development. Used by permission of the Society for Research in Child Development and the authors.

9. "The Process of Learning Parental and Sex-Role Identification" by David B. Lynn. From the *Journal of Marriage and the Family*, November, 1966, pp. 446–470. Copyright © 1966 by the National Council on Family Relations. Used by permission of the National Council on Family Relations and the author.

10. "Development of a Sense of Self-Identity in Children" by Carol J. Guardo and Janis Beebe Bohan. From *Child Development*, 1971, 42:1909–1921. Copyright © 1971 by the Society for Research in Child Development. Used by permission of the Society for Research in Child Development and the authors.

11. "Peer Interaction and the Behavioral Development of the Individual Child" by Willard W. Hartup. In E. Schopler and R. Reichler, eds., *Psychopathology and Child Development*. Copyright © 1976 by Plenum Publishing Corporation. Used by permission of Plenum Publishing Corporation.

12. "The Aftermath of Divorce" by E. Mavis Hetherington, Martha Cox, and Roger Cox. In National Association for the Education of Young Children, eds., *Mother/Child, Father/Child Relationships*. Copyright © 1978 by E. Mavis Hetherington, Martha Cox, and Roger Cox. Used by permission of the authors.

13. "Understanding the Young Adolescent" by David Elkind. From *Adolescence*, 1978, 13:127–134. Copyright © 1978 by Libra Publishers, Inc. Used by permission of Libra Publishers.

14. "Intrapsychic versus Cultural Explanations of the 'Fear of Success' Motive" by Lynn Monahan, Deanna Kuhn, and Phillip Shaver. From the *Journal of Personality and Social Psychology*, 1974, 29:60–64. Copyright © 1974 by the American Psychological Association. Used by permission of the American Psychological Association and the authors.

15. "The Problem of Ego Identity" by Erik H. Erikson. In Erikson, *Identity and the Life Cycle*. Copyright © 1959 by International Universities Press, Inc.

16. "Disturbance in the Self-Image at Adolescence" by Roberta G. Simmons, Florence Rosenberg, and Morris Rosenberg. From the *American Sociological Review*, 1973, 38:553–568. Copyright © 1973 by the American Sociological Association. Used by permission of the American Sociological Association and the authors.

17. "Changes in Family Relations at Puberty" by Laurence D. Steinberg. Based on a paper presented at the Society for Research in Child Development meetings, San Francisco, 1979. Copyright © 1977 by Laurence D. Steinberg.

18. "Breakups before Marriage: The End of 103 Affairs" by Charles T. Hill, Zick Rubin, and Letitia Anne Peplau. In George Levinger and Oliver C. Moles, eds., *Divorce and Separation*. Copyright © 1979 by the Society for the Psychological Study of Social Issues, Basic Books, Inc., New York. Used by permission of Basic Books.

19. "Transition to Parenthood" by Alice S. Rossi. From the *Journal of Marriage and the Family*, February 1968, pp. 26–39. Copyright © 1968 by the National Council on Family Relations. Used by permission of the National Council on Family Relations and the author.

20. "Work and Its Meaning" by Lillian Breslow Rubin. In Rubin, *Worlds of Pain*. Copyright © 1976 by Lillian Breslow Rubin, Basic Books, Inc., New York. Used by permission of Basic Books, Inc., and the author.

21. "The Subjective Experience of Middle Age" by Bernice L. Neugarten and Nancy Datan. From "The Middle Years," in the *American Handbook of Psychiatry* (2d ed.), volume 1, part 3, edited by Silvano Arieti. Copyright © 1974 by Basic Books, Inc., New York. Used by permission of Basic Books.

22. "The Midlife Transition: A Period in Adult Psychosocial Development" by Daniel J. Levinson. From *Psychiatry,* 1977, vol. 40. Copyright © 1977 by the William Alanson White Psychiatric Foundation, Inc. Used by permission of the William Alanson White Psychiatric Foundation and the author.

23. "Marital Satisfaction over the Family Life Cycle" by Boyd C. Rollins and Harold Feldman. From the *Journal of Marriage and the Family,* February, 1970. Copyright © 1970 by the National Council on Family Relations. Used by permission of the National Council on Family Relations and the authors.

24. "Primary Friends and Kin: A Study of the Associations of Middle-Class Couples" by Nicholas Babchuk. From *Social Forces,* 1965, 43:483–493. Copyright © 1965 by the University of North Carolina Press. Used by permission of the University of North Carolina Press. and the author.

25. "The Role of Grandparenthood" by Ivan F. Nye and Felix M. Berardo. In Nye and Berardo, *The Family: Its Structure and Interaction.* Copyright

© 1973 by Macmillan Publishing Co., Inc. Used by permission of Macmillan Publishing Co., Inc.

26. "Adjustment to Loss of Job at Retirement" by Robert C. Atchley. From *The International Journal of Aging and Human Development,* 1975, 6:17–27. Copyright © 1977 by Baywood Publishing Co., Inc. Used by permission of Baywood Publishing Co., Inc.

27. "Personality and Patterns of Aging" by Robert J. Havighurst. From *The Gerontologist,* 1968, 8:20–23. Copyright © 1968 by the Gerontological Society. Used by permission of the Gerontological Society.

28. "The Life Review: An Interpretation of Reminiscence in the Aged" by Robert N. Butler. From *Psychiatry,* 1963, 26:65–76. Copyright © 1963 by the William Alanson White Psychiatric Foundation, Inc. Used by permission of the William Alanson White Psychiatric Foundation, Inc. and the author.

29. "Interaction and Adaptation: Intimacy as a Critical Variable" by Marjorie Fiske Lowenthal and Clayton Haven. From the *American Sociological Review,* 1968, 33:20–30. Copyright © 1968 by the American Sociological Association. Used by permission of the American Sociological Association and the authors.

30. "The Meaning of Friendship in Widowhood" by Helena Znaniecki Lopata. In L. Troll, J. Israel, and K. Israel, eds. *Looking Ahead: A Woman's Guide to the Problems and Joys of Growing Older,* pp. 93–105. Copyright © 1977 by Prentice-Hall, Inc. Used by permission of Prentice-Hall, Inc.

Part I

Infancy
Mastery and Competencies

One

Competent Newborns
Tom Bower

Although our tendency is to think of infants as coming into the world in a relatively helpless state, research has shown us that newborns are far more competent than we had ever imagined. Ingenious experiments by infancy researchers such as Tom Bower have demonstrated that infants have, at birth, a highly functioning perceptual system which gives them the potential to "acquire new knowledge, skills, and competencies from the very moment of birth." In this selection, Bower discusses some of the problems and promises of conducting research on infants to discover just what they can—and cannot—do.

THE HUMAN NEWBORN is one of the most fascinating organisms that a psychologist can study. It is only after birth that psychological processes can begin; only after can success and failure, reward and punishment, begin to affect the development of the child. Before that point function and practice—for many theorists the motor forces in development—have no opportunity to modify the processes of growth that produce the neural structures in the brain that must underpin any capacities that are present at birth. The newborn infant is thus the natural focus of the age-old and continuing controversy between nativists and empiricists: that is, the argument over whether human knowledge is a natural endowment, like the structure of our hand, with differences in intellectual competence as genetically determined as differences in eye color, or is rather the product of behavior and experience in the world, with differences in intellectual capacity a function of differences in the quality of environmental exposure. An extreme proponent of the latter viewpoint would argue that the newborn with no exposure history behind him should therefore show no capacities at all, beyond the capacity to learn. Nativists would have to predict something very different.

3

To the casual eye the empiricists would seem to have the argument. The newborn seems extremely helpless, capable of little save eating, sleeping, and crying. But the casual eye would be in error. The human newborn is an extremely competent organism.

There are a number of precautions that one must take before trying to study newborn humans. The methods that we use to find out what newborns know must be adapted to characteristics of this fascinating organism. Obviously, no one would sit and ask a newborn questions and expect to get many answers. . . .

However, a newborn would not tell us in words how the world looked to him, for the obvious reason that newborns do not talk. If we want to find out what newborns know about the world, what they expect it to be like, we must rely on inferences from their behaviors.

We could not make many inferences about the newborns' knowledge of the world from observations of the spontaneous behavior of babies in Western culture. They have very few behaviors, and those that they have are not given much chance to appear in standard Western baby-care conditions. For example, newborns have some quite precise head and eye movements and hand and arm movements in their repertoire. However, when the baby is laid on his back, a standard examination position, these behaviors virtually disappear.

They disappear because the baby, in this position, must use his head and arms to hold himself in a stable position. If he picks up one arm to reach for something, he will roll in that direction. Even a head movement can result in a loss of postural equilibrium. The problem is compounded if the baby is wearing a large wad of diapers which tilt his weight up toward his head anyway. It follows that if one wants to use these head and eye and arm and hand behaviors to index the baby's knowledge of the world, one must place the baby in a position that allows him to move head and arms freely. In experiments in my lab we use makeshift arrangements of pillows to accomplish this, although more sophisticated baby chairs are available.

A second problem with newborns is not so easily solved. That is the problem of wakefulness. Newborns are awake for very brief periods, about six minutes at a time. At other times they may look awake, with eyes open and so on, and yet be functionally asleep. Without sophisticated apparatus to measure brain waves and other psychophysiological variables, it is hard to be sure that one is dealing with a fully awake baby. Lastly, if the baby is not fully awake, there seems to be no way of waking him up. Prechtle has presented evidence that the sleeping-waking cycles of newborns are under internal control and are relatively impervious to external events. Certainly, in my laboratory we have found that loud noises, flicks on the sole of the foot, and other techniques all fail to wake a baby who has gone off to sleep. In working with newborns, more than at any later stage of human development, the investigator must await the convenience of his subject.

The traditional starting point for an analysis of knowledge is the input that comes through the senses (sight, sound, smell, and touch, for example). That input does not seem capable of providing the information that adults are able to get out of it. Psychologists have assumed that the deficiencies in the sensory input were made up by learning and experience. But recent experiments with newborns must cast some doubt on this point of view.

Consider, for instance, the ability to localize an odor. Adults can localize odors, to right or left, with a fair degree of accuracy. This is a problem for psychology. Because there is no right or left in the nose, the right-left dimension of experience must be elaborated from other information. There are, of course, two sources of relevant information: the different intensities of odor at the two nostrils (an odor source on the right will stimulate the right nostril with greater intensity than it will the left); and the different times of arrival of the odor-producing molecules at the nostrils, a source on the right reaching the right nostril fractionally before it reaches the left. These time/intensity differences are used by the perceptual system to specify position to right and left. The structure that does this is present in newborns, who will turn smoothly away from "unpleasant" odor sources, indicating that they are capable of olfactory localization, as well as sharing adults' opinions of the pleasantness of some odors.

The auditory system poses similar problems. Adults can locate sounds to right and left, with great precision, although there is no right and left within the auditory system. Perception of position of a sound source is elaborated from time of arrival and intensity differences between the two ears, as well as patterns of reflection set up in the outer ear.

Michael Wertheimer demonstrated that within seconds of birth infants can use this information, turning their eyes correctly toward a sound source. This shows not only auditory localization but also auditory-visual coordination, an expectation that there will be something to be seen at a sound source.

My colleague Eric Aronson has confirmed the same basic point in more complicated experiments performed in Edinburgh. While an infant is in a special apparatus, he can see his mother through the soundproof glass screen, but can only hear her via the two speakers of a stereo system. The balance on the stereo can be adjusted to make the sound appear to come from straight ahead or any other position. If the mother speaks to her baby with the balance adjusted so that the heard voice appears to come from her seen mouth, the baby is quite happy. But if the heard voice and seen mouth do not coincide, very young infants manifest surprise and upset, indicating auditory localization, auditory-visual coordination, and, more surprisingly, an expectation that voices will come from mouths. This is an example of competence that seems to be lost with age. In lecture theaters or cinemas, for example, where heard voice and seen mouth are often in very different places, adults do not seem to be aware of the discrepancy.

The study of auditory localization is also important in that it allows us to begin to define the limits of innate structuring and to indicate just what perceptual capacities do require information from the environment. I said earlier that perception of position of a sound source is generated from time of arrival differences and intensity differences between the two ears. Consider how time of arrival differences function in auditory localization. A sound source which is straight ahead of an observer emits sound waves which reach both ears simultaneously. A sound source which is to the right emits sound waves which reach the right ear before the left ear. A sound source on the left emits sound waves which reach the left ear before the right ear. The further away a sound source is from the midline (to the left or right), the greater is the difference between time of arrival at the leading ear and time of arrival at the following ear. Human adults use these time of arrival differences to compute the precise position of a sound source.

Time of arrival differences do not depend simply on the position of the sound source—they also depend on the distance between the two ears. The further apart ears are, the greater will be the time of arrival difference produced by a given deviation from the midline plane. Newborn babies' ears are obviously much closer together than are the ears of human adults. As the baby grows, the ears become further apart as the head gets bigger. This raises the question of whether the newborn "knows" how far apart his ears are. In auditory localization the newborn shares some information with the adult: zero time of arrival differences signify that a sound source is in the midline plane (straight ahead or straight behind); when the right ear is stimulated before the left, the source is on the right and vice versa. These two items of information are invariant during growth. The information about the precise value to be attached to any given time of arrival difference, by contrast, changes during growth.

There are thus two types of information involved in auditory localization, one being dependent on growth, the other being independent. The baby's genetic blueprint could incorporate the growth independent information without too much trouble. It seems less likely that the growth dependent information could be so incorporated, especially since the growth of the head depends on the quality of nutrition available to the growing child.

Experimentation has shown this armchair speculation to be correct. We have done a variety of experiments on the precision of auditory localization of sound sources in various positions. In one set of experiments we put babies in a light-tight, dark room and then introduced noise-making objects at various positions around the baby. We compared his performance in this situation with his performance in a situation where the baby could see the object continuously, or else where he was shown the object briefly before the lights were switched out. When the noise-making object was straight ahead babies could localize it as or more accurately than they could localize a seen object. If the object was on the right they localized it to the right—if it was on the left they localized it to the left. However, the precision of their localization to the right or left in the dark was very poor indeed, much poorer than it was with either continuous or momentary visual information about the location of the object.

We conclude from these results that if the information required for a perceptual task is growth independent, it will be built (in the genes); if the information is growth dependent, however, it will have to be acquired through experience in the environment, it has to be learned. In our experiments, babies did not seem to acquire this growth dependent information until they were about seven months old. Not until that age did the precision of their reaching for a sound source off the midline attain the precision of their reaching for a sound source on the midline, or the precision of their reaching for a seen target in any location.

The hypothesis that growth invariant information is built-in and that growth dependent information must be acquired has proved useful in analyzing visual development and the coordination of hand and eye. The size of the eye changes as the baby grows, as does the distance between the eyes. Distance between the eyes is critical for three-dimensional vision. Our experiments indicate again that it is necessary for the baby to have some experience in the world before he can use this growth dependent information precisely. Similar problems crop up in reaching itself, where the length of arm changes drastically during growth. Very young babies do not seem to know precisely how long their arms are, reaching for objects that are out of reach and reaching past objects that are within reach.

Vision itself has attracted rather more attention than the other senses, reflecting its greater importance in normal function. Although the eye is more elaborate than the nose or ear, with a built-in structure to register right-left position, for example, there are many dimensions of visual experience that do not seem to be given directly in the visual input. For example, visual experience is clearly three dimensional, and the third dimension, distance, is clearly missing from the input to the eye. Beginning with Bishop Berkeley in the eighteenth century, empiricists have assumed that distance could only be gauged after man had learned to interpret clues for distance. It now seems that the interpretation of these signs, if this is what occurs, need not be learned.

With colleagues in America I discovered that infants can demonstrate adjustments to the distance of objects during the newborn period. For example, if one moves an object toward the face of a baby, he will execute a well-coordinated defensive movement, pulling back his head and bringing his hands and arms between himself and the object. The response occurs whether or not the object used is a real object or a purely visual simulation projected on a screen. The latter result indicates that the response is only elicited if the approaching objects come within a certain distance, which seems to be about 30 centimeters. Approach closer than this distance elicits defensive movements, even if the approaching object is small. Approach which stops further away than 30 centimeters will not elicit defensive movements, even if the approaching object is large.

This indicates that the infants are not responding simply to the size of the retinal image produced by the approaching object.

The fact that infants defend themselves against the approach of seen objects seems indicative, at least, of some expectation that seen objects are tangible. I have obtained more convincing evidence of such an expectation by presenting the infant with virtual objects, objects which are visible yet intangible. . . .

Newborn infants will reach out to touch seen objects. The behavior is very crude but does result in a high proportion of contacts if a real object is presented. What happens if a virtual object is presented? When the infant's hand reaches the seen location of the object there is no tactual input, since the object is intangible. If the infant expects seen objects to be tangible, this event should surprise him, as indeed it does. Infants presented with such objects react with extreme surprise and upset, indicating some degree of visual-tactual coordination.

In a recent experiment in my laboratory, Jane Dunkeld has shown that babies display a similar discrimination using the defensive response to an approaching object like those described above. If babies are presented with an approaching object, they defend themselves. If they are presented with an approaching aperture, which would pass them by harmlessly, they do not defend themselves. They thus discriminate between things and the spaces between things and can tell what type of information specifies the two types of display. The information they are picking up is quite subtle. Approaching objects cover up background texture while approaching apertures reveal it, a delicate discrimination that newborns can make.

None of the capacities described above would be at all obvious to a casual eye. They cannot be elicited from infants save under conditions which are somewhat special in our culture. The infant must be propped up and his head and arms must be free to move. This condition is not met if the baby is lying flat on his back, the most common position for baby observation in this culture. A newborn on his back uses head and arms to support himself in a stable position. Thus neither is free to engage in any of the indicator behaviors described above.

The experiments I have described so far indicate that new-born infants have a functioning perceptual system, with a striking degree of coordination between the senses. The last experiments on this line that I wish to mention provide the most striking instance of such coordinations: a coordination between vision and the baby's own body image. If a human adult seats himself or herself in front of a newborn infant (three weeks) and engages in any of a wide variety of face or finger movements, the infant will imitate the adult's movements. If the adult sticks his tongue out, the baby will retaliate. If the adult opens and closes his mouth repetitively, the baby will mimic the movements. Babies as young as this have no experience of mirrors, and yet they know they have a mouth and a tongue and can match the seen mouth and tongue of an adult to their own unseen mouth and tongue, a most striking demonstration of intersensory matching. The infant perceives a match, a similarity, between himself and adults, a capacity of immense significance for all development and particularly social development.

So far I have spoken of capacities that seem to be innate. I have not mentioned the one capacity that empiricists would maintain must be innate—the capacity to learn. The learning abilities of the newborn infant have been a major focus of research. Since new behaviors do appear at a great rate throughout infancy it is important, theoretically, to establish that learning can occur, and could, therefore, account for the emergent behaviors. Eight years ago it seemed that research had failed to demonstrate that the infant of less than six months could learn anything at all. This pessimistic conclusion has been totally overthrown to the extent that it has even been suggested that the newborn infant can learn better at that point in his development than he ever will again.

Psychologists assess learning with every simple paradigms involving rewards for specific activities. Increase in the rate of the rewarded activity is taken as a measure of learning. The problem is to make sure that the reward is really rewarding. If one can do this with infants they will readily demonstrate learning of a high order. Unfortunately, older infants are readily bored with most rewards. A few studies have used food, supplementary to normal diet, to elicit performance. Many more experimenters have used presentation of visual events to motivate the infant. The motivational problem seems simpler with very young babies. In the first few days of life one can demonstrate learning of a very high order. Neonate infants can learn not one but a *pair* of response-reward contingencies, requiring two different responses signaled by two different stimuli. For instance, Lew Lipsett, of Brown University, and I discovered separately that a three-day-old infant can learn to turn his head to the left to obtain reward when a bell sounds, and to the right when a buzzer sounds. He can learn the bell-left, buzzer-right discrimination in a few minutes. Having learned it, he can learn to reverse the discrimination if the experimenter reverses the contingencies, to go bell-right, buzzer-left, again very rapidly. The learning displayed here is possibly of a higher order than is ever displayed by an infra-human.

Born with a high native endowment, the human infant has the potential to acquire new knowledge, skills, and competences from the very moment of birth. The newborn thus forces us to a compromise between nativism and empiricism, possessing as he does enough capacity to make rapid learning possible.

The newborn may have done more than merely make us compromise, however; he may have forced us to reconsider the whole concept of development. A foundation that underpins the bulk of developmental psychology is the assumption that development is a continuous, step-by-step process, with older organisms knowing all that younger organisms know, plus some surplus. Appreciation of what happens to many of the capacities described above should make us reconsider this basic assumption. Many of the capacities of the newborn fade away in the course of development, some of them never to return. Neonate walking, a phenomenon I have not described, is a case in point. If newborns are held properly, they march along a solid surface in a most impressive manner. This capacity disappears at about the age of eight weeks. The reaching of newborns disappears at about the age of four weeks. Their ability, or perhaps willingness, to imitate goes at about the same time. I have already mentioned the loss of auditory-visual coordination in Aronson-type situations. That capacity goes in the first few months of life and, seemingly, does not come back, since, as I said, most adults simply do not notice auditory-visual discrepancy.

Similar problems occur with auditory localization. I mentioned that young babies will reach for a noise-making object in darkness, albeit somewhat inaccurately. In the course of development this coordination disappears. Babies over six months are very unlikely to reach for a noise-making object in darkness. In some cases the coordination does not reappear at all, so that the child becomes motorically paralyzed in total darkness. The simplest explanation of such losses of capacity would be that the capacities are simply not used and so atrophy from lack of use, as neural connections will.

That cannot be the whole explanation, since practice cannot stop the decline in walking or reaching. It can slow it but it cannot arrest it completely. The decline in hand-ear coordination cannot be stopped by practice either. Giving a baby a great deal of experience in darkness with noise-making objects will not arrest the drop in hand-ear coordination. Indeed, I have heard of one study of a blind baby who nonetheless lost the capacity right on schedule, despite a great deal of opportunity to use the coordination, and despite the fact that the baby was then left with no other sensory system to guide him in the world.

The loss of capacities under such conditions argues against simple lack of practice as an explanation. It seems rather as if some genetically determined developmental process is switching off the coordinations, despite environmental attempts to keep them switched on. The function of such a process is quite obscure. Its existence would seem to call into question the whole rationale for studies of very young infants. What is the point of studying them if their abilities are going to be switched off in the course of development? I have no ready answers to that question. However, there is some very exciting evidence that indicates that what happens to these capacities while they are present may specify the rate and course of later development.

Researchers have discovered that infants whose neonate walking is practiced will walk faster than infants who have not been so practiced. Infants whose reaching ability has been utilized in the neonate period produce true reaching earlier than infants who have not had such practice. Infants who have used hand-ear coordination extensively prior to its disappearance are more likely to recover the capacity than infants who have not had such practice, indicating that not only rate but also direction of development may be determined by the environment that these early capacities have. It is this possibility that makes the study of newborn capacities and their subsequent history in the world such an exciting area at the moment. Newborns are very similar to each other. Older children and adults are very different. It is possible that the effect of the environment on newborn capacities has a disproportionate influence on the development of such differences.

Two

Piaget's Observations as a Source of Hypotheses Concerning Motivation
J. McVicker Hunt

During the first two years of life, the infant's interactions with the environment become progressively more sophisticated. Whereas the newly born infant, at the beginning of what is called the *sensorimotor* period, deals with the world primarily through inborn reflexes, by the end of the first year, the infant is an active, motivated explorer of the environment. In this article. J. McVicker Hunt discusses development during the sensorimotor period (birth to about two years) in terms of the infant's motivation to explore the environment and acquire information. Not only do infants come into the world with important competencies and skills (see selection 1), they also may be naturally motivated to learn.

WHAT PIAGET HAS done in his studies of the early development of his own children is to observe and to describe the progressive changes in the structure of their interaction with their everyday circumstances. Among these circumstances, he has concerned himself almost exclusively with those in which the children were entirely comfortable and entirely satiated with food and water. The changes in the structure of their actions that he has described are those which took place in repeated encounters with various kinds of circumstances, and he has pointed out what the changes imply about such epistemological constructions as the permanence of objects and the grasp of causality, space, and time (Piaget 1937). Progressive change in structure, i.e., this epigenesis is marked by "landmarks of transition" which differentiate what Piaget call stages. Here I shall be concerned chiefly with the stages of the sensorimotor phase (Piaget 1936, 1937).

13

THE SENSORIMOTOR STAGES

In Sensorimotor Stage I, Piaget (1936) finds the neonate with some ready-made reflexive schemata. These schemata are sensorimotor structures that mediate behavior in each situation. Piaget's list of these schemata in the human neonate includes: sucking, looking, listening, vocalizing, grasping, and bodily motor activity. In this first stage, the duration of which can be expected to vary with the circumstances encountered (Piaget 1953), the human infant is, from the motivational standpoint, almost completely a responsive organism. It responds to homeostatic needs and to changes in the ongoing modalities of stimulation. In other words, that which the Russian investigators—whose work has been surveyed by Berlyne (1960) and by Razran (1961)—have termed the "orienting reflex" is present at birth. Note, however, that the instigators of looking and listening appear to be changes in ongoing stimulation rather than the impingement of a particular input.

Piaget's (1936) Stage II beings the coordination of these reflexive schemata. Changes in some characteristic of auditory stimulation come to evoke the looking. What is seen evokes reaching and, later, grasping. What is grasped evokes mouthing, and so forth. . . .

Piaget (1936) has conceived of such changes as examples of that invariant function which he calls *assimilation* and which he sets in contrast with *accommodation*. He conceives of the ready-made schema of looking as assimilating the changes in sound, when looking acquires the capacity to be evoked by them. Similarly, he sees the ready-made schema of grasping as assimilating the changes in the characteristics of light patterns, when grasping acquires the capacity to be evoked by the sight of an object, and so on. Of course, the sucking schema and the vocalizing schema—both of which are at first almost completely responsive to homeostatic need, as Irwin (1930) has shown—also become coordinated with the schemata of looking and listening in this same fashion.

One can properly say that the human child is responsive during these early stages precisely because it is homeostatic need, painful stimulation or, as the Russian students of the "orienting reflex" have shown (see Berlyne 1960; and Razran 1961), changes in the various characteristics of receptor input that evoke these schemata ready-made at birth. With experience, however, each schema gradually becomes coordinated with each other one via a process of sequential organization, or conditioning, or what Guthrie (1935) has termed contiguity learning. This coordination can be said to result from assimilation in the sense that each schema acquires the capacity to be evoked by the kinds of change in the circumstances of stimulation which formerly evoked only one of the other schemata. This coordination among the originally independent schemata is the major accomplishment of Stage II.

Piaget (1936) finds the onset of Stage III marked by the advent of a primitive "intention." It appears when the infant begins to manifest activity that is calculated to retain or elicit earlier forms of stimulation which have ceased. We are all familiar with this phenomenon. It is exemplified in the jouncing motions made by an infant who is on the knee of an adult who has been jouncing him but has stopped doing so. When the infant starts his jouncing, often when the attention of the adult has turned elsewhere, it is hard to avoid the impression that he anticipates the return of the adult's jouncing, wants it, and is purposefully endeavoring to get the adult to resume it. Such behavior may resemble the schmeikeling of a hungry cat around the ankles of someone who customarily has fed him. In the case of the jouncing infant, however, the goal of the activity is quite divorced from homeostatic need and painful stimulation. It is most likely to occur, in fact, when the infant is satiated for food and water and is entirely comfortable. Apparently the infant has come in the course of experience to like the jouncing for itself. I shall return to this point.

This advent of "intention" appears to me to be a major "landmark of transition" in the development of motivation, or at least of that kind which I like to call "intrinsic motivation" (Hunt 1963).

With this transition, the human child shifts from being an organism without initiative—an organism entirely responsive to various external circumstances from a motivational standpoint—to one that acts to get something the nature of which it anticipates in at least a primordial way.

Note that such purposive activities, which appear to be calculated to retain or elicit again some pattern of stimulation, occur only after an infant has already encountered that pattern of stimulation a number of times. The occurrence of such activities is then accompanied by an excited expression of attention which typically breaks into a smile and by other expressions of delight. Such activities are most commonly observed when the face of a human being moves into view, and they are then typically accompanied by a smile or a laugh.

In fact, Spitz (1946) has contended that smiling is a social response, strictly a social response, and one which registers delight from witnessing the facial constellation which has been associated with his previous satisfactions of food and comfort. But Piaget (1936) records many instances in which his children smiled for other than social stimulation. They smiled when the fringe on the hood of the bassinet shook, even though no person was in view. They smiled when familiar objects were hung from the hood. They smiled at the sight of their own hands coming before their eyes and Laurent, for example, was noted to smile after repeatedly finding his nose with his fingers. In each of these instances, however, the smile appeared only after the infant had encountered the circumstance repeatedly. From such evidence, Piaget gathered that the smile is a sign of recognition that comes with repeated encounters of a scene.

On the basis of such still uncertain evidence (which, by the way, I have been able to confirm to some degree with informal observations of my own). I should like to propose that the source of pleasure and excitement here is the emerging recognition. Presumably, once an infant has encountered a circumstance enough times to have established within his own brain somewhere a central process that can represent that circumstance, this process provides the basis for recognition, i.e., the central representational process can match and identify the input from the familiar situation. This recognition, I am suggesting, brings excited approaches to the source of the stimulation which we commonly call attention; brings the smile and even laughter; and brings activities to regain an encounter with the circumstances, once it has gone from perception.

At this stage the infant is, I am saying, no longer merely an organism that acts only in response to changes in the characteristics of ongoing stimulation, or to homeostatic need or painful stimulation or both. At this stage, he has begun to be an initiator of activity. The basis for his initiative resides in the relationship that exists between the information he already has stored from previous encounters with the circumstances in question and the information presently coming through his receptors. It is interesting to note, moreover, that it is at this stage that he begins to show distress at the absence of the familiar and recognized circumstances, situations, or patterns of stimulation. Now that the infant has at least the beginnings of intentions, he can be frustrated. This capacity to be frustrated increases as he begins to have plans when certain of his schemata become ends or goals, while other schemata take the role of instrumental means.

Such separation of ends from means marks the onset of Stage IV. This separation occurs when one of the infant's schemata, say that of grasping an object, becomes the end or goal and the other schemata, such as pulling, striking, pushing, or scotting, come to be used as instrumental means to achieve that goal. This is a primitive plan (see Miller, Galanter, and Pribram 1960). Piaget describes dozens of instances. In many of these the goal is to grasp some object or to see something only heard, but sometimes an action schema appears for a time to become an end in itself. Here is one example where the goal is grasping an object.

Observation 121: At age 8 months 20 days, Jacqueline tries to grasp the cigarette case which I present to her. I then slide it between the cross strings which attach her dolls to the hood [of the bassinette]. She tries to reach it directly. Not succeeding, she immediately looks at the strings which are not in her hands and of which she sees only the part in which the cigarette case is entangled. She looks in front of her, grasps the strings, pulls and shakes them. The cigarette case falls and she immediately grasps it. . . . Second experiment: same reactions but without trying to grasp the object so directly. (1936:215)

With this separation of means from ends comes a new emphasis on *accommodation*, that conceptual alterego of assimilation wherein the structure of the action is modified to fit the situation. From the standpoint of motivation, however, this new stage appears to be but an extension of Stage III. In it, reflecting the new emphasis on accommodation, the infant comes to show a more varied repertoire of spontaneous activities, and those increase in variety as the variations in the circumstances encountered call for accommodation. . . .

Piaget finds the onset of his Sensorimotor Stage V marked by the advent of an interest in novel circumstances, which I am suggesting as the second major "landmark of transition" in the epigenesis of intrinsic motivation. This third stage of motivational development occurs when the child begins to be intrigued by new or novel objects and by new or novel patterns of stimulation. With this new interest, moreover, comes active groping for new means in the achievement of goals. . . .

This fifth stage continues until the infant's activities come to be marked by the imaginal invention of new means, by imitation of models no longer present, by following objects through series of hidden displacements. All these characteristics attest the emergence of semi-autonomous central processes which can serve as representations of objects, events, and circumstances which have been encountered. The common word for these central representations is imagery. The advent of imagery brings Sensorimotor Stage VI, which at once terminates the sensorimotor phase of development and introduces the preconceptual phase. . . .

SOME HYPOTHESES SUGGESTED

The first hypothesis about the nature of motivation that Piaget's observations suggest to me is very general. They suggest that there must be some mechanism of motivation quite independent of painful stimulation and homeostatic need, those motivators usually thought of as the basis of our traditionally dominant theory. . . . Piaget's observations would strongly suggest that something in the infant's ordinary commerce with the environment through his receptor channels contains a basis for motivation. As I see it, Piaget's observations give us at least a first approximation of the epigenesis of what I now like to call "motivation inherent in information processing and action" or, as I have already noted, "intrinsic motivation" (Hunt 1963).

Second, Piaget's observations suggest an epigenesis of intrinsic motivation consisting of the stages that I have already indicated in my outline of his observations. In the first stage, beginning at birth, the infant is responsive from a motivational standpoint. In the second stage, beginning with the appearance of intentional activities calculated to prolong or elicit familiar spectacles and extending to plans in which one schema serves as the goal and others as the instrumental means, the child appears to be motivated by the desirability of those circumstances which have become familiar through repeated encounters. In the third stage, the infant becomes interested in what is new and novel within a complex of familiar circumstances.

Third, Piaget has stated one empirical principle of motivational development in the aphorism that, "the more a child has seen and heard, the more he wants to see and hear" (1936:277). The marked apathy and retardation commonly observed in orphanage infants is certainly consonant with this principle. You will recall that Dennis (1960) has found that 60 percent of the children of a Teheran orphanage, where the circumstances of stimulation are especially homogeneous, still do not sit up alone by two years of age, and 85 percent of the children of this orphanage still fail to walk alone at four years of age. Piaget's observations suggest an explanation of this empirical principle which is particularly interesting to me. It is suggested, by the observation that infants seek and enjoy objects and patterns with which they have become familiar through repeated encounters, that even at three, or four, or five months of age the interest which a child has in his environment is a function of the variety of these objects and patterns which he has already encountered during his earlier weeks of life. Moreover, it is suggested, by the observation that children (beginning near the end of their first year) become interested in novel variations of those objects and patterns with which they are already familiar, that the interest and responsiveness of the toddler is a function of the variety of objects and patterns with which he has had an opportunity to become familiar during the first year of his life.

Such considerations also suggest that the retardation in the development of both motor and intellectual functions in infants growing up in orphanages may well be the consequence of retardation in intrinsic motivation, deriving from having encountered insufficient variation in circumstances. This hypothesis is readily subject to experimental investigation with animals. Perhaps we should also be using the orphanages of the world as a source of evidence while the conditions still exist that permit them to be sources of evidence. In addition this line of still somewhat speculative theorizing suggests that the times between these "landmarks of transition" should be examined in relation to the interest parents have in their children at the various stages of development. Lois Murphy (1944) has already reported observations indicating that varying rates of development at various stages may well be a function of the manner in which parents handle their infants at those stages of their first two or three years of life. . . .

Fourth, Piaget's observations suggest an interesting conception both of the motivation for such repetitive autogenic activities as babbling and hand-watching and of the motivational basis for imitation and its role in socialization. Inasmuch as Piaget has observed his children come and go through various body wigglings each time he stopped making a peculiar noise that he had repeated in their presence, it is only a short step of inference to see a child's repetition of his own vocalization patterns as examples of the same effort to continue or to elicit interesting spectacles, albeit self-produced, vocal spectacles. If the sight of a repeated pattern becomes a source of recognitive delight, why should this not be the motivational basis for the hand-watching so commonly observed in infants at about four months of age? You will recall that in winter-born Jacqueline (Piaget 1936)—who spent a large share of her waking, daylight hours all bundled up so that she could be outdoors in the sun—this hand-watching schema did not appear until after age six months. You will also remember that in late-spring-born Lorenz, with whom Piaget began his observations and "little experiments" almost immediately following birth, this hand-watching schema made its appearance at age two months. . . .

Fifth, Piaget's observations also suggest an explanation of the fact that children can easily be separated from their familiar parents and from familiar surroundings during the first few months of their lives, that is, in the period before recognition has been thoroughly established. They explain in turn the severity of the separation anxiety, or perhaps a better term would be the *separation grief,* characteristic of children during the last third of the first year and during the second year.

SUMMARY

I have tried in these remarks to remind you of the nature of Piaget's observations. I have said that they strongly suggest to me that there must be a mechanism of motivation independent of the painful stimulation and homeostatic need commonly used to explain why organisms are active in what has come to be the traditionally dominant conception of motivation. I have said that Piaget's observations provide at least a first approximation of an outline of the epigenesis of what I am calling the "motivation inherent in information processing and action" or "intrinsic motivation." I have noted Piaget's empirical generalization that "the more a child has seen and heard, the more he wants to see and hear."

I have attempted to explain not only this empirical generalization, but also several other commonly observed behaviors of infants and their role in development, with the notion that once children have been exposed to a given pattern of stimulation enough times to make it familiar, the emerging recognition of the pattern brings pleasure that motivates an effort to retain or reelicit the pattern. I have further attempted to elaborate this explanation with the notion that, after a pattern has continued to be familiar for a time, it is variation in that pattern that brings pleasure and the effort to find that variation in either the child's own activities or in external stimulation. One can consider none of these suggestions established, but they constitute hypotheses for investigation.

REFERENCES

Berlyne, D. E. 1960. *Conflict, Arousal, and Curiosity.* New York: McGraw-Hill.

Dennis, W. 1960. Causes of retardation among institutional children. *J. Genet. Psychol.* 96:47–59.

Guthrie, E. R. 1935. *The Psychology of Learning.* New York: Harper.

Hunt, J. McV. 1963. Motivation inherent in information processing and action. In O. J. Harvey, ed., *Motivation and Social Organization: Cognitive Determinants.* New York: Ronald.

Irwin, O. C. 1930. The amount and nature of activities of new-born infants under constant external stimulating conditions during the first ten days of life. *Genet. Psychol. Monogr.* 8:1–92.

Murphy, Lois B. 1944. Childhood experience in relation to personality. In J. McV. Hunt, ed., *Personality and the Behavior Disorders.* New York: Ronald.

Piaget, J. 1936. *The Origins of Intelligence in Children.* Trans. by Margaret Cook. New York: International Universities Press, 1952.

—— 1937. *The Construction of Reality in the Child.* Trans. by Margaret Cook. New York: Basic Books, 1954.

—— 1953. Remarks about himself. In J. M. Tanner and Bärbel Inhelder, eds., *Discussions on Child Development.* Proceedings of the First Meeting of the World Health Organization Study Group on the Psychological Development of the Child. New York: International Universities Press.

Razran, G. 1961. The observable unconscious and the inferable conscious in current Soviet psychophysiology: interoceptive conditioning, semantic conditioning, and the orienting reflex. *Psychol. Rev.* 68:81–147.

Spitz, R. A. 1946. The smiling response: a contribution to the ontogenesis of social relations. *Genet. Psychol. Monogr.* 34:67–125.

Infancy
Identity and the Self

Three

Mirror-Image Stimulation and Self-Recognition in Infancy

Jeanne Brooks-Gunn and Michael Lewis

One of the first developments in the domain of identity and self is the ability to recognize oneself in a mirror. When do infants show signs of recognizing themselves? In this cleverly designed experiment, Brooks-Gunn and Lewis examine self-recognition in infants ranging in age from nine months to two years. Although some signs of recognition were seen at all ages, it was more apparent among the older infants. This suggests that the ability to recognize oneself is something that emerges over time.

THREE QUESTIONS are of interest in the study of the origins of self in man: How does the infant come to know itself? What does it mean for the infant to know itself? How can the infant's knowledge be known to us? The above questions assume that the young child does have a concept of self, that this knowledge is active and changing, and that early self-knowledge may be known by others. The present investigation is an attempt to study empirically the development of self in infancy.

One way of studying the development of the self is to find an aspect of self-knowledge that is easy to define and to observe. One

This paper was presented at the Biennial meetings of the Society for Research in Child Development, Denver, April 1975 and was revised slightly in 1979. Major portions are to be found in M. Lewis & J. Brooks-Gunn, *Social Cognition and the Acquisition of Self* (New York: Plenum Press 1979).

The research reported in this paper was supported by a grant from the National Institute of Mental Health (MH-248449-02). We wish to thank Christine Brim, Sherrill Lord, and Al Rogers for their assistance in data collection and analysis.

25

such aspect is visual self-recognition. Visual self-recognition, or at least facial recognition, is almost universal in our society due to repeated exposure to mirrors and to pictures. Though very young children have been thought not to recognize themselves, little systematic study has actually been undertaken. The present study attempts to remedy this lack of empirical investigation by exploring the development of visual self-recognition in mirrors.

There is a long history of the use of mirrors for self-appraisal. Humans have been known to look at themselves in mirror-like surfaces for at least 3,000 years (Swallow 1937). The phylogenetic and ontogenetic histories are also rich. The only adults who have difficulty recognizing themselves visually are persons suffering from certain CNS dysfunctions, severely mentally retarded children and adults, and some psychotic patients (Frenkel 1964; Harris 1977; Shentoub, Soulairac, and Rustin 1954; Wittreich 1959). Phylogenetically, self-recognition does not occur in any species except the great apes, including man (Gallup 1968, 1973).

What do we know about self-recognition and mirror behavior in the young? There are at least four sources of information: mother's reports, diaries on infant development, infant intelligence tests, and experimental studies. Mothers often report that their infants enjoy mirror play, and they sometimes use a mirror to soothe a fussy infant. Social scientists have also made this observation and have realized that the mirror may be used to measure self-recognition. Almost a hundred years ago, Preyer (1893) and Darwin (1877) both observed that mirror-image stimulation elicited great interest and curiosity in their children. Darwin observed what he thought was self-recognition in his nine-month-old son, who would turn toward a mirror when his name was called. Of course, today this observation would be attributed to learning, not self-recognition.

Infant intelligence test developers were also interested in early mirror behavior, and all of them included mirror items in their scales (Bayley 1969; Buhler 1930; Cattell 1940; Griffiths 1954; Gesell 1928, 1934). However, only Buhler and Gesell included mirror-related behavior that occurred after the first year of life and that may be indicative of self-recognition. Gesell was the only infant test developer to comment on the existence of self, believing that self-recognition did not occur in the first two years of life. One preschool test developer did see self-recognition as an ability present in the young child. In the standardization of the Merrill-Palmer Scale, Stutsman (1931) found that two-thirds of the two-year-olds recognized or labeled themselves upon seeing themselves in a mirror. Mirror self-recognition was included in the test as a second-year item.

Even more surprising than the early test developers' lack of interest in self-recognition is the experimental psychologists' lack of interest. Only a few investigators have systematically studied the development of self-recognition in terms of mirror-image stimulation. Dixon (1957), Amsterdam (1972), and Bertenthal and Fischer (1978) observed different aged infants in front of a mirror and postulated age-related stages of mirror behavior. Dixon outlined four stages—(1) "Mother," (2) "Playmate," (3) "Who do dat when I do dat?" and (4) "Coy." In the "Mother" stage, the infant enjoys observing another's movement in the mirror; in the "Playmate" stage, the infant responds playfully to his or her own image (as if it were a peer); in the "Who do dat when I do dat?" stage, the infant is interested in observing his or her actions; and in the "Coy" stage, the infant acts coy, shy, or fearful in front of the mirror. Dixon believed the "Coy" stage to be indicative of self-recognition.

Amsterdam's stages are similar: the first involves social responding to the mirror (smiling at, vocalizing to, approaching, and patting the mirror), the second is the beginning of self-awareness (acting self-conscious, fearful, coy, and averting one's gaze), and third involves self-recognition (self-directed rather than other-directed behavior). The social stage is prevalent from six to twelve months of age, the transitional stage from twelve to eighteen months of age, and the self-recognition stage from twenty to twenty-four months of age. Amsterdam also reports little overlap between stages.

These two studies, although interesting, are inadequate for several reasons. First, the sample sizes are small (five infants were seen longitudinally by Dixon and four infants were seen each month through the first two years of life by Amsterdam). Second, there were procedural difficulties in that the infants could see their mothers and the observer in the mirror in both studies, and the infants were confined in a playpen for 7½ minutes in the Amsterdam study. These procedural difficulties present two problems, the first having to do with the ecological validity of the situation (observers are not present in the home and infants do not usually engage in mirror play in a playpen), and the second having to do with the reflection to which the infant is responding (are the infants smiling and vocalizing to their image, their mothers, or the observers?). Third, the behavioral criteria for the existence of the stages could be more .rigorously defined and the observations could be better standardized. The present study was designed with these problems in mind.

EXPERIMENTAL DESIGN AND PROCEDURE

Design

In the present study, infants' reactions to mirrors were observed using an ingenious technique independently developed by Gallup (1968, 1970, 1973) and by Amsterdam (1968, 1972). Sixteen infants (eight males and eight females) in each of six age groups were observed in the following mirror situation. Each infant was first placed in front of a large mirror and observed (No Rouge Condition). Then, a dot of rouge was placed on the infant's nose by the mother and the infant was observed in front of the mirror (Rouge 1 Condition). Then, the experimenter applied a dot of rouge to the mother's nose and the infant's reaction to the mother's marked face was noted (Mother Rouge Condition). After seeing the mother, the infant was placed in front of the mirror for a third time (Rouge 2 Condition). The four conditions were labeled No Rouge, Rouge 1, Mother Rouge, and Rouge 2.

Comparisons

The study was designed so that three comparisons could be made. However, only the data regarding the first comparison will be presented (see Lewis and Brooks-Gunn, 1979, for a more complete description and analysis of the study). Briefly, the comparisons are as follows.

1. *The Effect of Rouge Application on Self-Directed Behavior.*

Each infant's responses in the No Rouge and the Rouge 1 Conditions were compared to see whether the presence of a mark on infants' faces affects their subsequent behavior in front of a mirror. Using an unmarked condition as a baseline is essential since the incidence of spontaneous face- or nose-directed behavior has never been systematically observed. Amsterdam's (1968, 1972) study may be criticized for not including an unmarked condition.

2. *A Comparison of Mark Recognition on the Mother and on the Self.*

Infants' ability to recognize a mark on their mother's nose as well as their own were also compared. Although the Mother Rouge Condition was designed primarily as a training technique, the salience of a mark on the face in general is of importance, since the effects of facial distortions per se, with the exception of scrambled features, have not been studied. Perhaps a small dot of rouge on the nose would not be discovered by a certain-aged infant, whereas a larger distortion would. This comparison provides us with information on recognition of facial distortion independent of recognition of self.

3. *The Effect of Maternal Mark Recognition on Subsequent Self-Directed Behavior.*

The mother mark condition may be seen as a training technique as well as a salience test, since recognition of rouge on the mother's nose might facilitate recognition of the rouge on oneself. Thus, the incidence of self-directed behavior in the Rouge 1 and Rouge 2 Conditions may be compared in terms of whether or not maternal mark recognition occurred. The crucial comparison involves the infants who do not touch their noses in the Rouge 1 Condition but do tough their mother's noses. Are these infants more likely to recognize the mark in the Rouge 2 Condition than infants who do not touch their mother's noses?

Procedure

Infants were brought to the laboratory by their mothers and greeted by the experimenter. Each infant was given approximately five minutes to "warmup" while the procedure was explained to the mother. The following instructions were given:

We will observe your child's behavior in front of a mirror. Mirror play is of interest since we believe that it is related to self-recognition. To test this idea, the following procedure has been designed and will be used. First, I will take you and your child to another room where a large mirror has been placed on the floor. There is a camera behind the mirror which will record your child's mirror play. After I have left the room, please enourage your child to go to the mirror. You may place him in front of the mirror, tell him to go the mirror, sit beside the mirror, or place the small chair in front of the mirror. *Do not sit in front of the mirror yourself.*

After I have observed your infant in front of the mirror from the observation window, I will reenter the mirror room. I will give you a cloth which has rouge on it. Please wipe your child's nose with the rouge so that his nose is noticeably red. Tell the child that you are wiping his/her face because it is dirty. *Do not mention* the child's nose or the *rouge*. I will then leave the room and go to the observation window. I will knock on the window three or four times. Each time, try to get your child to look at himself in the mirror. Again, do not mention the child's nose or the rouge and do not sit where your image would reflect in the mirror.

After your infant has observed himself, I will come back into the room and apply rouge to *your* nose. Then I will leave and knock two different times on the observation window. At the first knock, get your child to look at you by sitting next to him or by picking him up. *Talk to him but do not mention your nose, his nose, or the rouge.* On the second knock, please have your child go to the mirror alone again. Are there any questions?

The infant was then taken to the mirror room, which was pleasantly decorated and which contained two chairs, a table, and wall posters. There were no toys in the room so as not to detract from the mirror. The mirror itself was a 46 cm. × 89 cm. one-way mirror mounted in a large 1.22 m. × 2.44 m. piece of plywood. The entire structure fit between two of the walls in the room and formed a triangle with the walls and the corner of the room. A camera was placed inside this triangle and was covered with a black cloth to reduce the amount of light on the back side of the one-way mirror.

The *E* left the room, only reentering prior to the rouge applications. The mother applied the rouge to the infant, the *E* to the mother. The ruse of wiping the face was effective, as only one of the 96 infants touched his nose immediately after the rouge was applied and before looking in the mirror.

Measures

We were interested in a large number of behavioral responses. These may be classified as follows: facial expressions, vocalizations, attention, mirror-directed behaviors, self-directed behaviors, and imitative behaviors. Although all of these behaviors will not be discussed here, the listing gives an idea of the large range of behaviors exhibited toward the mirror by the infants.

Observer Reliability

The infants' responses to the mirror conditions were video-taped, and behavioral coding was done from these videotapes. Eleven subjects were randomly selected for reliability purposes and their tapes coded by two observers. Observer reliability was calculated by the following formula: number of agreements/number of agreements and disagreements. The percentage of agreements between the two observers for the eleven subjects was quite high. The percentages ranged from 88 to 100 for mirror-directed, self-directed, and imitative behaviors. Agreements were also high for smiling (93 percent) and frowning (100 percent), but were somewhat lower for the concentrate expression (70 percent).

Interobserver reliability for attention was calculated differently. Since number of looks was coded, the percentage of agreements may be calculated by using the number of looks as the denominator, the number of agreements as the numerator. Using this method, the in-terobserver reliability for number of looks was 84 percent.

RESULTS

The present paper deals with the effect of rouge application on sub-sequent mirror behavior. We were interested in whether or not certain behaviors changed as a function of the mark application and as a function of age. That is, were some behavioral changes more likely to occur in the older than in the younger infants?

Five different behavioral categories were examined: these are attention, facial expression, mirror-directed behavior, self-directed behavior, and imitative behavior. The absence or presence of each behavior for each condition was coded, so that data are in the form of percentage of subjects. Percentage difference scores for each age group between the No Rouge and Rouge 1 Conditions are presented in the following figures. These percentages reflect both the magnitude and the direction of change. Positive scores indicate that more infants exhibited a specific behavior in the Rouge than in the No Rouge Condition.

Nose-Directed Behavior

The central issue of the study was whether or not the infants noticed the rouge. Nose- (or mark-) directed behavior increased dramatically from the No Rouge to the Rouge 1 Condition, occurring only twice in the former and 30 times in the latter condition. Nose-touching was also highly related to age. None of the 9- or 12-, three of the 15-, four of the 18-, eleven of the 21-, and twelve of the 24-month-olds exhibited mark-directed behavior ($X^2(5) = 40.05$, $p < .001$). This increase was a monotonically increasing function.

Figure 3.1 presents pictorially the age of subject $\times$ stimulus condition interaction, which was highly significant ($X^2(2) = 32.63$, $p < .001$). The increase in nose-touching, of course, only occurred in the older age groups. Thus, nose-touching was directly related to the rouge application, as it rarely occurred in the base condition. In addition, the age of the infant strongly dictated whether or not mark-directed behavior occurred. Would other behaviors exhibit such trends?

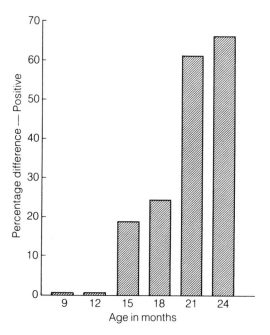

Figure 3.1. Nose-directed behavior: Percentage difference between the No Rouge and Rouge 1 Conditions by age.

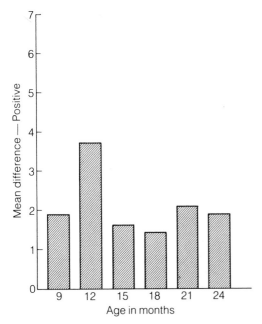

Figure 3.2. Total number of looks: Mean difference score between the No Rouge and Rouge 1 Conditions by age.

Attention

The number of looks directed toward the mirror increased dramatically after the rouge application. There were one and one-half times as many looks in the Rouge 1 as in the No Rouge Condition (F (1, 66) = 38.92, p < .001). The increase was found for all age groups, as is shown in figure 3.2.

Another measure of attention or interest is the concentrate expression. The concentrate face has only recently been considered noteworthy (Brooks and Lewis 1976; Sroufe, Matas, and Waters 1974) and in fact has often been considered a negative response (Lewis and Brooks 1974; Morgan and Ricciuti 1969; Scarr and Salapatek 1970). However, the characteristics of this expression (open mouth shaped like an "O" or an ellipse, eyes wide open, eyebrows raised) are related to cardiac deceleration (Sroufe, Matas, and Waters 1974), which suggests that this expression may be associated with attention and interest rather than negative affect.

In fact, the concentrate facial expression data complement the looking data well. More infants exhibited a concentrate expression in the marked than in the unmarked condition ($X^2 = 21.95$, $p < .001$). Figure 3.3 presents the percentage difference scores for the concentrate expression by age. As can be seen, there were no age differences, as the increase in concentrate expression was found for all age groups. Thus, increased interest and attention were associated with the mark application. This was true for all age groups, even the younger ones who did not exhibit mark recognition.

Affect

There were no stimulus condition or age group differences with respect to pleasant and negative vocalizations or facial expressions. All age groups were highly likely to exhibit positive affect and were not likely to exhibit negative affect in either condition.

Mirror-Directed Behavior

In general, mirror-directed behaviors were not affected by the rouge application for the total sample, with one interesting exception.

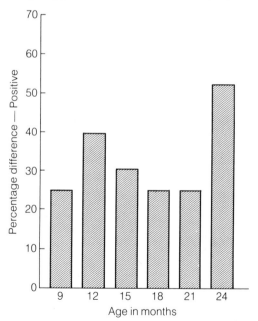

Figure 3.3. Concentrate expression: Percentage difference between the No Rouge and Rouge 1 Conditions by age.

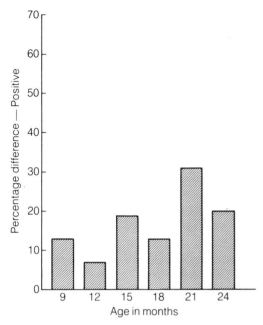

Figure 3.4. Touches image in the mirror: Percentage difference between the No Rouge and Rouge 1 Conditions by age.

More infants touched their own *image* in the mirror during the marked than the unmarked condition ($X^2 = 4.36$, $p < .05$). This stimulus effect was influenced by age. As can be seen in figure 3.4, the older infants were more likely to touch the image in the marked than the unmarked condition than were the younger ones ($X^2(2) = 7.87$, $p < .02$).

Imitation
        Three imitative categories were coded: these are (1) bouncing, waving, and clapping, (2) making faces, and (3) acting silly or coy. Only acting silly or coy was affected by the mark application. As is shown in figure 3.5, the older but not the younger infants were more likely to act silly or coy in the marked condition ($X^2(2) = 6.84$, $p < .05$).

Body-Directed Behavior
        Body-directed as well as nose-directed behavior might also increase in the marked condition, as recognition of the mark might

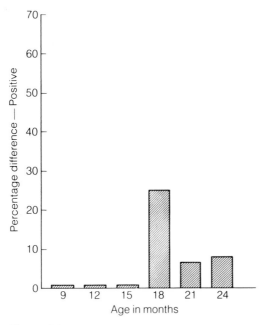

Figure 3.5. Acts silly or coy: Percentage difference between the No Rouge and Rouge 1 Conditions by age.

result in more general exploration of the body as well as of the mark itself. In fact, such an increase did occur, as the difference scores in figure 3.6 indicate (X^2 (2) = 6.84, $p < .05$). As is shown in figure 3.6, the increase in body-directed behavior was seen in all age groups, even in those which did not exhibit mark recognition.

DISCUSSION

Mark Recognition

   The central issue of the study was whether or not the rouge application would affect infants' responses to the mirror and would result in mark recognition. Recognition of the rouge was clearly demonstrated, as thirty infants touched, wiped, or verbally referred to the marked nose immediately following the rouge application. Only two infants touched the nose prior to the mark application, indicating that

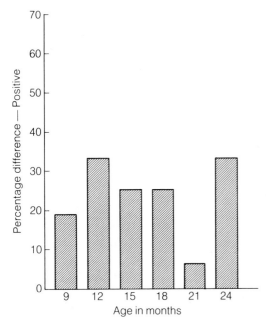

Figure 3.6. Body-directed behavior: Percentage difference between the No Rouge and Rouge 1 Conditions by age.

spontaneous nose-touching did not affect our results. Striking developmental trends were found as none of the 9–12-, one-quarter of the 15–18-, and three-quarters of the 21–24-month-olds touched their noses.

How do our findings relate to those of others? Amsterdam (1968, 1972) and Bertenthal and Fischer (1978) also used a mark-on-the-face technique. Mark recognition occurred earlier (15 versus 18 months) and more frequently (32 percent versus 17 percent) in the present study than in Amsterdam's (1968), while the trends in our study and that of Bertenthal and Fischer (1978) were quite similar.

What might account for the earlier and more frequent self-recognition in our sample? First, procedural differences may have affected the results, since Amsterdam's infants were confined to a playpen for a long period of time, had had their clothes removed immediately before testing, were seen in conjunction with a medical checkup, and had an observer present. Thus, Amsterdam's situation may have been more stressful than ours. Second, the social class composition of the samples were somewhat different, with Amsterdam's sample being more heterogeneous than ours. However, there were no effects of social class with respect to self-recognition within either sample.

Mark Recognition and Other Response Patterns

The Rouge application also affected the expression of other behaviors. The infants looked at themselves more often, were more likely to touch their image in the mirror, act silly and coy, touch their own body and to exhibit a concentrate expression in the marked than in the unmarked condition. There were important age differences in the expression of differential responding. The increase in number of looks, concentrate expression, and body-directed behavior occurred in all age groups, while acting silly or coy, touching one's nose, and touching one's image only increased in the older infants.

From which measures are we to infer self-awareness? If we use self-directed behavior as an index, then all age groups were aware of themselves. If only mark-directed behavior is used, then self-awareness is age related. The problem of measurement is not easily resolved, especially since age-related physical coordination may be affecting the ability to direct behavior visually toward the mark. Pointing requires the infant to direct attention to that which is being pointed at, not the pointer, and this ability seems to have a developmental course similar to role-taking (Flavell 1974a, 1974b; Masangkay et al. 1974). Young infants may not be able to distinguish between the act of pointing and the receiver of the point (Lewis and Brooks-Gunn 1979). Regardless of the measure used, by 15 months of age, infants are able to exhibit behaviors clearly indicative of self-awareness. That infants as young as 9 months may show self-awareness is an unanswered question and awaits further measurement refinement.

The Development of Self-Recognition:
Does a Stage Theory Explain the Findings?

Both Amsterdam (1968, 1972) and Dixon (1957) suggested a stage theory to explain infants' responses to mirrors. In both conceptualizations, a stage seems to be defined by the appearance of certain behavior clusters which are superseded by new behavior clusters as a function of age. Each behavior cluster is predominant at a certain age, slowly disappearing as the new cluster enters the child's repertoire. Amsterdam outlines three stages: social responding or playmate stage, transitional or self-conscious stage, and self-recognition or marked-directed stage. She suggests that self-recognition is not evident before a certain age and that it is preceded by transitional behaviors, such as coy or self-conscious behaviors.

Our data lend little support to such a stage theory. Most of the behaviors observed were not seen exclusively at one age. Some behaviors remained constant over the six age groups, some gradually increased, and some decreased with age. All of the infants were interested in the mirror and interacted with it. Smiling, touching, and pleasant vocalizations, which had been considered evidence of a playmate stage by Dixon and Amsterdam, were exhibited by over three-fifths of all age groups. Other behaviors were affected by age. For example, mark-directed behavior and acting silly or coy increased with age. However, none of these behaviors formed discrete clusters or were exhibited in only one age group.

The fact that the rouge application affected responding also weakens the stage theory concept. Even the youngest infants, none of whom exhibited mark recognition, reacted differently prior to and following the rouge application. This suggests that mark recognition is not a discontinuous phenomenon.

SUMMARY

The present study allows for a preliminary look at the process by which infants come to recognize themselves. It is preliminary in that mirrors are only one medium in which recognition occurs; infants experience their image in pictorial form as well as in home movies and videotapes. By examining responses to different representational forms, the development of self-recognition can be more precisely delineated, an endeavor that is reported by Lewis and Brooks-Gunn (1979).

In addition, the use of mirrors, pictures, and videotapes allows for the separation of the various dimensions of self-recognition. Contingency is a feature of mirrors but not pictures, movement of videotapes and mirrors but not pictures, and self-other comparisons of pictures and videotapes but not mirrors. In *Social Cognition and the Acquisition of Self*, we outline the developmental trends for the various dimensions of self-recognition.

REFERENCES

Amsterdam, B. K. 1968. Mirror behavior in children under two years of age. Ph.D. dissertation. Chapel Hill: University of North Carolina.
—— 1972. Mirror self-image reactions before age two. *Developmental Psychology* 5:297–305.
Bayley, N. 1969. *Bayley Scales of Infant Development*. New York: Psychological Corporation.
Bertenthal, B. I. and K. W. Fischer. 1978. Development of self-recognition in the infant. *Developmental Psychology* 14:44–50.
Brooks, J. and M. Lewis. 1976. Infants' responses to strangers: midget, adult, and child. *Child Development* 47:323–332.
Bühler, C. 1930. *The First Year of Life*. New York: John Day.
Cattell, P. 1940. *The Measurement of Intelligence of Infants and Young Children*. New York: Psychological Corporation.
Darwin, C. 1877. A biographical sketch of an infant. *Mind* 2:285–294.
Dixon, J. C. 1957. Development of self recognition. *Journal of Genetic Psychology* 91:251–256.
Flavell, J. H. 1974a. The development of inferences about others. In T. Mischel, ed., *Understanding Other Persons*. Totowa, N.J.: Rowman and Littlefield.
—— 1974b. The genesis of our understanding of persons: psychological studies. In T. Mischel, ed., *Understanding Other Persons*. Totowa, N.J.: Rowman and Littlefield.
Frenkel, R. E. 1964. Psychotherapeutic reconstruction of the traumatic amnesic period by the mirror image projective technique. *Journal of Existentialism* 17:77–96.
Gallup, G. G., Jr. 1968. Mirror-image stimulation. *Psychological Bulletin* 70:782–793.
—— 1970. Chimpanzees: self-recognition. *Science* 167:86–87.
—— 1973. Towards an operational definition of self-awareness. Paper presented at the Ninth International Congress of Anthropological and Ethnological Sciences, Chicago.
Gesell, A. 1928. *Infancy and Human Growth*. New York: MacMillan.
Gesell, A. and H. Thompson. 1934. *Infant Behavior: Its Genesis and Growth*. New York: McGraw Hill.

Griffiths, R. 1954. *The Abilities of Babies*. London: University of London Press.

Harris, L. P. 1977. Self-recognition among institutionalized profoundly retarded males. A replication. *Bulletin of the Psychonomic Society* 9:43–44.

Lewis, M. and J. Brooks-Gunn. 1974. Self, other and fear: infants' reactions to people. In M. Lewis and L. Rosenblum, eds., *The Origins of Fear: The Origins of Behavior*, vol. 2. New York: Wiley, pp. 195–227.

—— 1979. *Social Cognition and the Acquisition of Self*. New York: Plenum.

Masangkay, Z. S., K. A. McCluskey, C. W. McIntyre, J. Sims-Knight, B. E. Vaughn, and J. H. Flavell. 1974. The early development of inferences about the visual percepts of others. *Child Development* 45(2):357–366.

Morgan, G. A. and H. N. Ricciuti. 1969. Infants' responses to strangers during the first year. In B. M. Foss, ed., *Determinants of Infant Behavior*, vol. 4. London: Methuen Press.

Preyer, W. 1893. *Mind of the Child*, vol. 2. *Development of the Intellect*. New York: Appleton.

Scarr, S. and P. Salapatek. 1970. Patterns of fear development during infancy. *Merrill-Palmer Quarterly* 16:53–90.

Shentoub, S. A., A. Soulairac, and E. Rustin. 1954. Comportement de l'enfant arrière devant le miroir. *Enfance* 7:333–340.

Sroufe, L. A., L. Matas, and E. Waters. 1974. Determinants of emotional expression in infancy. In M. Lewis and L. Rosenblum, eds., *The Origins of Fear: The Origins of Behavior*, vol. 2. New York: Wiley.

Stutsman, R. 1931. *Mental Measurement of Preschool Children*. Yonkers, N.Y.: World Book.

Swallow, R. W. 1937. *Ancient Chinese Bronze Mirrors*. Peking, China: Henri Vetch.

Wittreich, W. 1959. Visual perception and personality. *Scientific American* 200:56–60.

Four

Characteristics of the Individual Child's Behavioral Responses to the Environment

Stella Chess, Alexander Thomas, and Herbert Birch

An important aspect of an individual's identity is his or her personality—that person's unique way of responding to the environment. Although much of an individual's personality is shaped by important persons and events over the course of a lifetime, there is a great deal of evidence that important personality and temperament differences among individuals are already present at birth. Over the past twenty-five years, Stella Chess, Alexander Thomas, and Herbert Birch have been studying these early, and apparently innate, personality differences. In this selection, the authors suggest that very young infants do not all respond in identical ways to similar behaviors by their parents.

RECENT YEARS have witnessed a great increase in our knowledge of the part played by the character of the family constellation in influencing both healthy and unhealthy child development. Such factors as maternal overprotection, maternal rejection, loss of the mother, withdrawal, or a restrictive, punitive approach by the father, and sibling rivalry have been found to be associated with disturbances in the psychological development of the child. In addition, environmental influences outside the immediate family unit, such as the school situation and the more general social and cultural environment, have been and continue to be areas of much productive investigation.

Paralleling the various specific studies in this direction, there

42

has also developed an approach to child care practices based on an exclusive focus on the importance of environmental factors. For the young child this has involved an emphasis on the role of the mother, and secondarily the father and siblings. For the older child there has been an additional focus on the school and various play situations. Parents and teachers have been flooded with advice as to the optimal approaches to children calculated to ensure the healthiest psychological development.

Implicit in most of these formulations has been the assumption that each child will react in the same way to any specific approach by the parent, whether in feeding, toilet training, discipline, or any other area of functioning. A child care practice which has a favorable effect on some children is assumed to be desirable for all; a practice which has unfavorable effects on some children is considered undesirable for all. Where a particular child care practice appears to have varying effects on different children, explanations for these deviations tend to be given in terms of the presence of counteracting influences in the mother, father, or sibling relationship.

This emphasis on the role of the mother has stimulated a host of psychological studies in which normal child development and psychological upsets in the young infant, and behavior disorders, psychosomatic disturbances, delinquency, and schizophrenia in the older infant and child, are examined with the only variable considered being that of the behavior and attitude of the mother. Differences in the child are considered to be entirely the result of the influence of the mother, or the combination of this central maternal force with other environmental factors.

Even where a particular child care formula appears to emphasize an attention to the individual characteristics of the child, there is often the same underlying assumption that all babies will react similarly to the same situation.

For example, self-demand feeding in the infant is recommended because it permits the child's individual needs and preferences to be expressed, and not smothered by an arbitrarily imposed rigid feeding schedule. However, this self-demand technique to be successful necessitates that the infant be able to express his hunger in a clear and definite fashion. His hungry cry has to be differentiated from other crying, has to be delayed to the point where the stomach is almost or entirely emptied, and not delayed too long beyond this point. A recommendation for an unrestricted self-demand feeding technique for all babies assumes that all infants have such an identifiable and defined response to hunger.

As another example, a permissive approach to child discipline is advocated on the same basis, namely, that it allows the child to express and develop his own individual needs. Here again, to be successful permissiveness necessitates that the child respond by spontaneously developing constructive, socially desirable patterns of functioning. A recommendation for unrestricted permissiveness in all cases assumes that all children will respond in this way.

Fifty years ago infants were fed strictly by the clock. A number of babies did badly on this regime, and a new rule of self-demand feeding was recommended. It is now found that some infants do not do well on this approach, and that a modified or rigid schedule is more successful in such cases. This same swinging of the pendulum has been apparent in the rules recommended for handling the crying baby, for toilet training, for teaching rules of behavior to the two-year-old, and for teaching reading to the six-year-old. In each case it has not been found possible to find a formulation that is successful with all children.

The history of the child care field in recent years suggests, therefore, that the attempt to find rules of management applicable to all children may not be possible. However, efforts continue to be directed toward finding new sets of rules in the persistent hope of developing a universally applicable generalization. . . .

INDIVIDUAL DIFFERENCES IN REACTIVITY

The present paper reports a test of the hypothesis that the effect of various child care practices is determined not only by what the mother feels and does but also by the specific intrinsic pattern of reactions which characterizes the individual child, by presenting data derived from an ongoing longitudinal study of 85 children. We believe that the data indicate that the individual specific reaction pattern appears in the first few months of life, persists in a stable form thereafter, and significantly influences the nature of the child's response to all environmental events, including child care practices.

Our research parallels the theme of a number of studies which have reported individual differences in reactivity in the infant which appear to be of a nonexperiential character. Fries and Woolf (1953) found striking differences in a group of young infants in the functioning of the neuromuscular system, and classified them as quiet, moderately active, and active congenital activity types. They suggested that the individual activity type might persist and influence personality development. Escalona, Leitch and their coworkers (1952) have reported a detailed cross-sectional study of the behavior of a group of 128 infants of varying ages. Definite individual differences in various activities of the infant such as sleep, feeding, and response to sensory stimulation, were found (Escalona 1952). A longitudinal study of individual infant development has been reported from the Yale Child Center (Senn 1954). Gesell (1948) and Ilg and Ames (1955:64–65) have reported observations on individual behavior patterns which they considered innate. Biochemical aspects of individual differences have been approached by Williams (1956) and Mirsky (1953). A number of workers have also begun to gather data on various specific physiological aspects of individual differences in autonomic functioning of infants (Bridge and Reiser 1957; Grossman and Greenberg 1957; Richmond and Lustman 1955).

Subjects and Methods

The study reported here is a longitudinal analysis of the behavioral development of a series of 85 children. Whereas seven have been followed impressionistically from birth for a period of eight to fourteen years, the remainder have been followed on a systematic protocol basis from the first few months of life. The latter detailed series of 78 cases has been accumulated since March 1956.

Developmental information has been gathered on two levels: detailed history taking from the parents and/or other persons involved in the care of the child, and direct observations by outside observers of cross-sectional segments of the child's behavior.

The history is first taken when the baby is two to three months old, and then at three-month intervals in the first year and at longer intervals thereafter. The direct observations have included: a two- to three-hour period of observation of the child in his usual home setting by one or two persons unfamiliar with the behavioral history; observations of the child's behavior by his pediatrician; and the observation of behavior in standard test situations, such as the Gesell Development Test in infants, and a psychometric test in older children. In several cases, it has been possible to obtain meaningful supplemental and confirmatory data from close friends of the child's family.

The methodology for data collection and some of the concepts for data evaluation have been described in two previous reports (Thomas and Chess 1957a, 1957b). The approach to gathering information involves two principles:

1. The behavior of the child is to be described in objective, noninterpretative terms. Whenever a judgment is made by the parent or another observer, such as that the baby "liked" a certain food or that he "couldn't stand it," the insistence is on a specific, step-by-step description of the actual behavior.

2. The various specific items of the child's behavior are considered to indicate responses to stimuli from the external or internal environment. The attempt is made to describe the behavior in its relationship to the specific situation in which it occurs. Special attention is paid to the response on first contact with a new stimulus, such as a new food, the first experience with the bath, the first hypodermic injection, the first haircut, first contacts with various social, learning, and disciplining situations, and so on. Also recorded in detail is the response to the same stimulus on subsequent exposures until there appears a consistent, long-term pattern of response. Inquiry is made into any subsequent situations that change or modify this consistent response, either temporarily or permanently, and into the nature of this modification. Attention is given to the various alterations in the child's routine or environment as representing changes in the effectice stimuli acting on the child. Changes may include such items as attempts to alter the baby's feeding or sleep schedule, illness, a move to different living quarters, the arrival of a new baby in the family, and so on.

The histories taken from the parents have proven to be a very productive approach to the gathering of detailed behavioral data, including the sequence of responses of the child on different occasions. The raw data for determining the baby's specific reaction pattern must include not only the details of his specific behavior in various areas at any one time, but also the consistency or changes in his responses on exposure to the same situations at other times. Such information can be obtained only from individuals involved in the day-to-day care of the child.

The families in the study represent a fairly homogeneous middle-class group. On a conscious level, the mothers uniformly accept their major responsibility as being the care of the baby, with their emphasis placed on the importance of satisfying the baby's needs. They have shown great variation in their responses to the specific, objective questions regarding the child's behavior. Some mothers give factual, objective answers, while others repeatedly become involved in subjective, interpretive evaluations of the meaning of the behavior. Some are concise, others ramble. Some can give a wealth of detail regarding various items of their children's behavior, while others give much sparser material. Some mothers are overtly preoccupied with making value judgments on the behavioral development of their babies and on their own child care practices, while still others show little or no open concern in these areas. We have found, however, that a history-taking technique which insists on answers in terms of specific, factual details of behavior can elicit a great deal of objective data even from mothers who are subjective, interpretive, and preoccupied with value judgments, and who ramble or tend to be sparse in the content of their replies. The material obtained in this way at any one interview can be broken down into an average number of approximately 70 separate items of behavior which can be scored and categorized by methods of content analysis.

Results

The details of developing reliable scoring procedures will be presented elsewhere. At this point it can be stated that the behavior reported in the parent interviews could be inductively analyzed in terms of nine separate categories that could be scored at all of the ages studied. These categories are:

1. Activity—Passivity
2. Regularity—Irregularity
3. Intense Reactor—Mild Reactor
4. Approacher—Withdrawer
5. Adaptive—Nonadaptive
6. High Threshold—Low Threshold
7. Positive Mood—Negative Mood
8. Selective—Nonselective
9. Distractibility—Nondistractibility

Each category refers to a characteristic mode of functioning as reported for the child, and behavioral features of normal life situations were analyzed for each category in separate independent analyses of the protocols.

1. Activity—Passivity refers to the magnitude of the motor component present in a given child's function and to his diurnal proportion of active and inactive periods. Therefore, protocol data on mobility during bathing, eating, playing, dressing, and handling, as well as information concerning the sleep-wake cycle, reaching, crawling, walking, eating, and play patterns are used in scoring for this functional category.

2. Regular—Irregular refers to the predictability and rhythmicity or unpredictability and arhythmicity of function and can be analyzed in relation to the sleep-wake cycle, hunger, elimination, appetite, and demand cycles.

3. Intense—Mild refers to the quality of response and its vigor, independent of its direction. A negative response may be either mild or intense as can a positive response. Responses to stimuli, to pre-elimination tension, to hunger, to repletion, to new foods, to attempts at control, to restraint, to dressing and diapering all provide scorable items for this category.

4. Approach—Withdrawal represents a category of responses to new things, be they people, foods, or toys. In it the behaviors reported are scored for the nature of initial responses.

5. Adaptive—Nonadaptive again refers to responses to new or altered situations. However, in this category one is not concerned with the nature of the initial responses, but with the ease with which such responses are modified in desired directions.

6. High Threshold—Low Threshold is an omnibus category in which subcategories are concerned with (a) sensory threshold, (b) responses to environmental objects, and (c) social responsiveness.

7. Positive Mood—Negative Mood represents a category in which degrees of pleasure-pain, joy-crying, friendliness-unfriendliness are rated.

8. Selectivity—Nonselectivity refers to the definition of function and to the difficulty with which such an established direction of functioning can be altered. It is a composite of persistence and attention span.

9. Distractibility—Nondistractibility refers to the ease with

which new peripheral stimuli can divert the child from an ongoing activity.

In developing scale divisions for the scoring of behaviors in each category listed, preliminary investigation revealed that a 3-point scale represented the maximum refinement of scale divisions that could be effectively utilized. Difficulties in more extended scales (4- and 5-point scales) developed in connection with the intermediate values. It was found that if several judgmental positions were used to separate the extremes—e.g., Active and Passive—from one another, marked unreliability in intermediate placement occurred. Consequently, it was decided to use a single intermediate scale position and therefore to use a 3-point scale for the scoring of behavioral items under each category.

On blind analysis by two independent judges, 90 percent agreement was obtained in the categorical scoring of 22 serially obtained cases. This method of analysis also permitted the segmental direct observations to be compared with the data reported in the parent interview. The agreement between direct observation and parent report was reliable to between the .05 and .01 level of confidence for a series of 19 sequentially selected cases.

The direct observations, on the other hand, are not a substitute, but rather a complement, to the detailed histories obtained from the parents. The observations only catch a variable segment of the child's activity and cannot, unless they were repeated at very frequent intervals, record the sequences of development and change in reactions that may occur even from day to day.

Analysis of consistencies in each of the children from one age period to another with regard to the persistence of pattern of behavioral responses confirms the hypothesis of the existence of an intrinsic reaction type. When the interperiod scores in the nine categories described above are subjected to chi-square agreement-disagreement contingency analyses, it is found that the interperiod agreements are of such magnitude that the likelihood that they could have occurred by change alone or as a function of sampling is less than 1 percent. This conclusion is based upon the statistical analysis of the characteristics of the first 32 children in our population who achieved the age of twenty-four months. Consistencies in reaction type, therefore, are found to exist for at least the first two years of life. Statistical analysis of the remaining children in our population is in progress, and age levels beyond twenty-four months are being examined in the same fashion as these data become available.

In addition, a qualitative evaluation of each record in the entire series supports the quantitative analysis by clearly indicating the persistence at successive age levels of the characteristic reaction pattern already evident at two to three months of age.

Our data thus far do not permit a definite answer to the question of whether these reaction patterns are of an inborn character, or formed under the influence of environmental factors in the first few months of life, or the resultant of the interaction of these two factors. Our records do include a great deal of material on parental attitudes and functioning in the majority of our children and we are hopeful that a projected detailed analysis of these data will provide some basis for answering this question of genesis. However, our impressions from the evidence incline us to the opinion that these patterns are not experientially determined, but are of an intrinsic character. What may be necessary to give a clear-cut answer may be a study of a series of babies which will include detailed, longitudinal observations of behavior from birth through the first two months of life.

APPROACH TO CHILD CARE PRACTICES

To return to the question raised at the beginning of the paper, namely, the need to individualize child care practices, the behavioral data in this series of 85 children point up this issue very emphatically. Our consistent finding has been that the response of the child to the parental approach in various areas and at different ages has been determined not only by the attitude and behavior of the parent but also by the child's own specific reaction pattern. This has been true in sleeping, feeding, and bathing in the young infant; in discipline, play, and responses to people in the older infant; in toilet training, and in social and learning situations in the older child. In all these areas there have been children who showed differing responses with parents whose approaches have been similar. There have also been children who showed similar responses with parents whose approaches have differed. . . .

The majority of the children in our series show patterns involving clear-cut, consistent reactions of moderate and graded intensity, with the ability to form long-term responses in various areas quickly but without rigidity. It is our impression that within broad limits such babies could do well with differing child-care practices, as long as the parents are consistent in their approach. It is with these children that even differing concepts of child care can be formulated and applied successfully. What is perhaps more important, however, is that a substantial minority of babies show other types of reaction patterns which do not permit favorable responses to differing parental approaches. Various illustrations of this phenomenon have been given above under feeding, discipline, learning, etc. It is for such children that individualized approaches become important. However, for an individualized approach to be effective, it must be based on knowledge of the particular child's specific reaction pattern. Without an attempt to define these patterns, formulations such as "Approach the child as an individual" and "It is important to know your own baby's individuality" tend to become platitudes.

As a final point, if the child's patterns of behavior and emotional responses are determined not only by parental attitudes and other environmental factors but also by his own specific reaction pattern, then caution must be imposed on the common tendency to assume that disturbances in a child are necessarily the exclusive results of unhealthy parental functioning. This may be true in some cases, but in others it is quite possible that parental functioning which is undesirable for that particular child might be constructive and desirable for another child. What has impressed us over the years has been the destructive impact on parents of the prevalent concept that they are the exclusive determinant of disturbances in the child's development. Many of the mothers of problem children develop enormous guilt feelings due to the assumption that they must necessarily be solely responsible for their children's emotional difficulties. With this guilt comes anxiety, defensiveness, increased pressures on the children, and even hostility toward them for "exposing" the mother's inadequacy by their disturbed behavior. Bruch (1954) has described very vividly the destructive consequences of the prevalent "illusion of omnipotence" in parent education. She quotes Dr. McIntosh, former president of Barnard College, regarding the impact on college students of the current approaches to child care and motherhood:

They have acquired fixed opinions, from child psychology courses and from reading, which set their future responsibilities in a most terrifying light . . . all the experts seem to be saying to them: "Even the most innocent-appearing act or a carelessly spoken word may harm a child or damage his future happiness. You hurt them by comparing them to other children; you hurt them by not comparing them and praising them for being special; you hurt them by being too affectionate to them and by not being affectionate enough." (1954:727).

The present study reported in this paper does suggest that the concept of the omnipotent role of the parent in the shaping of the child is indeed a destructive illusion.

SUMMARY

A report is given of a longitudinal study of 85 children with the purpose of gathering data regarding the characteristics of the individual child's behavioral responses to environmental stimuli. The finding of specific, stable reaction patterns in each child is reported, with the evidence thus far available indicating a nonexperiential origin. The methodology of the study and a scheme of classifications of the patterns are given briefly. The importance of these reaction patterns in influencing the effect of child care practices in the individual child is discussed.

REFERENCES

Bridger, W. and M. Reiser. 1957. Preliminary report of psychophysiological studies of the neonate. Presented at New York Divisional meeting, American Psychiatric Assoc., Nov. 17.

Bruch, H. 1954. Parent education or the illusion of omnipotence. *Am. J. Orthopsychiatry* 24:723–732.

Chess, S., H. Birch, and A. Thomas. A methodology for the study of individual behavioral characteristics in children. Manuscript.

Escalona, S., M. Leitch, et al. 1952. Early phases of personality development. *Monogr. Soc. Res. Child Developm.* 17:2.

Escalona, S. 1952. Emotional development in the first year of life. In *Problems of Infancy and Childhood: Sixth Conference,* New York: Josiah Macy, Jr. Foundation, pp. 11–92.

Fries, M. E. and P. J. Woolf. 1953. Some hypotheses on the role of the congenital activity type in personality development. *The Psychoanalytic Study of the Child* 8:48. New York: International Universities Press.

Gesell, A. 1948. The doctrine of development in child guidance. In L. G. Lowrey, ed., *Orthopsychiatry 1923–1948*. New York: American Orthopsychiatric Association, pp. 211–216.

Grossman, H. J. and N. H. Greenberg. 1957. Psychosomatic differentiation in infancy. *Psychosom. Med.* 19:293.

Ilg, F. L. and L. B. Ames. 1955. *Child Behavior*. New York: Harper.

Mirsky, I. A. 1953. Psychoanalysis and the biological sciences. In *Twenty Years of Psychoanalysis*. New York; Norton, pp. 155–176.

Richmond, J. B. and S. L. Lustman. 1955. Autonomic function in the neonate. *Psychosom. Med.* 17:269.

Senn, M. J. E. 1954. Research on personality development of the child. *Proc. Ass. Res. Nerv. Ment. Dis.*, 33:232–238.

Thomas, A. and S. Chess. 1957a. An approach to the study of sources of individual differences in child behavior. *J. Clin. Exp. Psychopath.* 18:347.

—— 1957b. Intrinsic reaction patterns: a factor in personality development. Presented at New York Divisional Meeting, American Psychiatric Assoc., November 15.

Williams, R. V. 1956. *Biochemical Individuality*. New York: Wiley.

Infancy
Relations with Others

Attachment, Exploration, and Separation: Illustrated by the Behavior of One-Year-Olds in a Strange Situation

Mary D. Salter Ainsworth and Silvia M. Bell

Few concepts have been as important in influencing our understanding of interpersonal development during infancy as *attachment*. An attachment is "an affectional tie that one person or animal forms between himself and another specific one—a tie that binds them together in space and endures over time." In this selection, Ainsworth and Bell discuss the significance of the infant's attachment to the parent and describe a procedure through which the nature of this attachment can be assessed. Understanding the infant-parent attachment helps us to understand the infant's reactions to parents, to strangers, and to novel situations.

IT IS THE PURPOSE of this paper to highlight some distinctive features of the ethological-evolutionary concept of attachment by citing reports of the interactions between the infant's attachment behavior and other behaviors mentioned above; to illustrate these interactions by a report of the behavior of one-year-olds in a strange situation; and to note parallels between strange-situation behavior and behavior reported in other relevant observational, clinical, and experimental contexts.

Let us begin with some definitions and key concepts distinctive of the ethological-evolutionary viewpoint, as proposed by Bowlby (1958, 1969) and Ainsworth (1964, 1967, 1969). *An attachment* may be defined as an affectional tie that one person or animal forms between himself and another specific one—a tie that binds them together in space and endures over time. The behavioral hallmark of attachment is seeking to gain and to maintain a certain degree of proximity to the object of attachment, which ranges from close physical contact under some circumstances to interaction or communication across some distance under other circumstances. *Attachment behaviors* are behaviors which promote proximity or contact. In the human infant these include active proximity- and contact-seeking behaviors such as approaching, following, and clinging, and signaling behaviors such as smiling, crying, and calling.

The very young infant displays attachment (proximity-promoting) behaviors such as crying, sucking, rooting, and smiling, despite the fact that he is insufficiently discriminating to direct them differentially to a specific person. These initial behaviors indicate a genetic bias toward becoming attached, since they can be demonstrated to be either activated or terminated most effectively by stimuli which, in the environment of evolutionary adaptedness, are most likely to stem from human sources. When these behaviors, supplemented by other active proximity-seeking behaviors which emerge later—presumably through a process of learning in the course of mother-infant interaction—become organized hierarchically and directed actively and specifically toward the mother, the infant may be described as having become attached to her.

The intensity of attachment behavior may be heightened or diminished by situational conditions, but, once an attachment has been formed, it cannot be viewed as vanishing during periods when attachment behavior is not evident. Therefore, it seems necessary to view attachment as an organization of behavioral systems which has an internal, structural portion that endures throughout periods when none of the component attachment behaviors have been activated.

Viewed in the context of evolutionary theory, infant-mother attachment may be seen to fulfill significant biological functions, that is, functions that promote species survival. The long, helpless infancy of the human species occasions grave risks. For the species to have survived, the infant has required protection during this period of defenselessness. It is inferred, therefore, that the genetic code makes provision for infant behaviors which have the usual (although not necessarily invariable) outcome of bringing infant and mother together.

Exploratory behavior is equally significant from an evolutionary point of view. As Hamburg (1968) has pointed out, a prolonged infancy would miss its adaptive mark if there were not also provisions in the genetic code which lead the infant to be interested in the novel features of his environment—to venture forth, to explore, and to learn. The implication is that the genetic biases in a species which can adapt to a wide range of environmental variations provide for a balance in infant behaviors (and in reciprocal maternal behaviors) between those which lead the infant away from the mother and promote exploration and acquisition of knowledge of the properties of the physical and social environment, and those which draw mother and infant together and promote the protection and nurturance that the mother can provide. . . .

Naturalistic studies of the attachment-exploration balance are very time consuming; the interaction between the two sets of behaviors must be observed over a wide range of situations. A short-cut alternative is to utilize a controlled strange or unfamiliar situation in which the child, with and without his mother, is exposed to stressful episodes of different kinds. So powerful is this technique in evoking behavioral changes that it is likely to be used with increasing frequency in studies of mother-infant interaction. The ethological-evolutionary view of the attachment-exploration balance is a useful model to use when planning and when interpreting the findings of strange-situation studies. . . .

The strange-situation procedure provides more than an opportunity to observe how exploratory behavior is affected by mother-present, mother-absent, or other conditions. It is a laboratory microcosm in which a wide range of behaviors pertinent to attachment and to its balance with exploratory behavior may be elicited. Attachment behaviors may be seen as complicated by "negative" behaviors, such as avoidance and aggression. And yet, since the laboratory situation provides but a very small sample of mother-infant interaction, strange-situation findings are not self-interpreting. Perception of the implications of the behaviors that occur in it is facilitated by reference to the findings of other studies—naturalistic, clinical, and experimental. For this reason the ensuing report of a strange-situation study is presented as a useful *illustration* of the shifting balance between exploratory and attachment behavior implicit in the ethological-evolutionary view of attachment. The discussion which follows the presentation refers to relevant findings of other studies. The propositions offered in conclusion comprehend these other relevant considerations as well as the findings of the illustrative strange-situation study.

THE STRANGE SITUATION

In the course of a longitudinal, naturalistic investigation of infant-mother attachment during the first year of life, there was little opportunity in the home environment to observe the balance of attachment and exploratory behaviors under conditions of novelty and alarm. Therefore, a laboratory situation was devised as a test situation to which the Ss were introduced when nearly one year old. It was desired to observe the extent to which the infant could use his mother as a secure base from which he could explore a strange environment, with fear of the strange kept in abeyance by her presence. It was also intended to observe the extent to which attachment behavior might gain ascendancy over exploratory behavior under conditions of alarm introduced by the entrance of a stranger and under conditions of separation from and reunion with the mother.

Method

Subjects. The 56 Ss were family-reared infants of white, middle-class parents, who were originally contacted through pediatricians in private practice. One subsample of 23 Ss, who had been observed longitudinally from birth onward, were observed in the strange situation when fifty-one weeks old. The second subsample of 33 Ss, studied in the context of an independent project (Bell 1970), were observed when forty-nine weeks old.

Procedure. The strange situation was comprised of eight episodes which followed in a standard order for all subjects. The situation was designed to be novel enough to elicit exploratory behavior, and yet not so strange that it would evoke fear and heighten attachment behavior at the outset. The approach of the stranger was gradual, so that any fear of her could be attributed to unfamiliarity rather than to abrupt, alarming behavior. The episodes were arranged so that the less disturbing ones came first. Finally, the situation as a whole was intended to be no more disturbing than those an infant was likely to encounter in his ordinary life experience. A summarized account of the procedure has been given elsewhere (Ainsworth and Wittig 1969) but will be reviewed here.

The experimental room was furnished—not bare—but so arranged that there was a 9 × 9-foot square of clear floor space, marked off into 16 squares to facilitate recording of location and locomotion. At one end of the room was a child's chair heaped with and surrounded by toys. Near the other end of the room on one side was a chair for the mother, and on the opposite side, near the door, a chair for the stranger. The baby was put down in the middle of the base of the triangle formed by the three chairs and left free to move where he wished. Both the mother and the female stranger were instructed in advance as to the roles they were to play.

In summary, the eight episodes of the situation are as follows:

Episode 1 (M, B, O). Mother (M), accompanied by an observer (O), carried the baby (B) into the room, and then O left.

Episode 2 (M, B). M put B down in the specified place, then sat quietly in her chair, participating only if B sought her attention. Duration 3 minutes.

Episode 3 (S, M, B). A stranger (S) entered, sat quietly for 1 minute, conversed with M for 1 minute, and then gradually approached B, showing him a toy. At the end of the third minute M left the room unobtrusively.

Episode 4 (S, B). If B was happily engaged in play, S was nonparticipant. If he was inactive, she tried to interest him in the toys. If he was distressed, she tried to distract him or to comfort him. If he could not be comforted, the episode was curtailed—otherwise it lasted 3 minutes.

Episode 5 (M, B). M entered, paused in the doorway to give B an opportunity to mobilize a spontaneous response to her. S then left unobtrusively. What M did next was not specified—except that she was told that after B was again settled in play with the toys she was to leave again, after pausing to say "bye-bye." (Duration of episode undetermined.)

Episode 6 (B alone). The baby was left alone for 3 minutes, unless he was so distressed that the episode had to be curtailed.

Episode 7 (S, B). S entered and behaved as in episode 4 for 3 minutes, unless distress prompted curtailment. (Ainsworth and Wittig 1969 planned a somewhat different procedure for episode 7, which was attempted for the first 14 *Ss* but, as it turned out, approximated the simpler procedure reported here, which was used for the remaining *Ss*.)

Episode 8 (M, B). M returned, S left, and after the reunion had been observed, the situation was terminated.

The behavior of the *S*s was observed from an adjoining room through a one-way vision window. Two observers dictated continuous narrative accounts into a dual channel tape recorder which also picked up the click of a timer every 15 seconds. . . .

The narrative record yielded two types of measure. A frequency measure was used for three forms of exploratory behavior—locomotor, manipulatory, and visual—and for crying. . . .

The second measure was based upon detailed coding of behaviors in which the contingencies of the mother's or stranger's behavior had to be taken into consideration:

Proximity- and contact-seeking behaviors include active, effective behaviors such as approaching and clambering up, active gestures such as reaching or leaning, intention movements such as partial approaches, and vocal signals including "directed" cries.

Contact-maintaining behaviors pertain to the situation after the baby has gained contact, either through his own initiative or otherwise. They include: clinging, embracing, clutching, and holding on; resisting release by intensified clinging or, if contact is lost, by turning back and reaching, or clambering back up; and protesting release vocally.

Proximity- and interaction-avoiding behaviors pertain to a situation which ordinarily elicits approach, greeting, or at least watching or interaction across a distance, as when an adult entered, or tried to engage the baby's attention. Such behaviors include ignoring the adult, pointedly avoiding looking at her, looking away, turning away, or moving away.

Contact- and interaction-resisting behaviors included angry, ambivalent attempts to push away, hit, or kick the adult who seeks to make contact, squirming to get down having been picked up, or throwing away or pushing away the toys through which the adult attempts to mediate her interventions. More diffuse manifestations are angry screaming, throwing self about, throwing self down, kicking the floor, pouting, cranky fussing, or petulance.

These four classes of behavior were scored for interaction with the mother in episodes 2, 3, 5, and 8, and for interaction with the stranger in episodes 3, 4, and 7.

Search behavior was scored for the separation episodes 4, 6, and 7. These behaviors include: following the mother to the door, trying to open the door, banging on the door, remaining oriented to the door or glancing at it, going to the mother's empty chair or simply looking at it. Such behaviors imply that the infant is searching for the absent mother either actively or by orienting to the last place in which she was seen (the door in most cases) or to the place associated with her in the strange situation (her chair.)

In scoring these five classes of behavior, the score was influenced by the following features: the strength of the behavior, its frequency, duration, and latency, and by the type of behavior itself—with active behavior being considered stronger than signaling. Detailed instructions for scoring these behaviors as well as for coding the frequency measures are provided elsewhere.

Findings

The findings to be reported here are of behaviors characteristic of the sample as a whole. Individual differences were conspicuous, instructive, and significantly correlated with other variables.

Exploratory Behavior. Figure 5.1 shows how three forms of exploratory behavior vary in successive episodes from 2 through 7. There is a sharp decline in all forms of exploratory behavior from episode 2 when the baby was alone with his mother to episode 3 when the stranger was present also. (This and all other interepisode differences reported here are significant at the .01 level or better, as tested by the binomial test, unless noted otherwise.) Exploration remains depressed through episode 4 when the baby was left with the stranger. Visual and manipulatory exploration (visual at the .02 level) recover significantly in episode 5, aided by the mother's attempts to interest the baby again in play, although similar efforts by the stranger in episodes 4 and 7 were ineffective. Visual and manipulatory exploration decline again in episode 6 after the mother departs for a second time, leaving the baby alone. All forms of exploratory behavior decline to their lowest point in episode 7 after the stranger had returned but while the mother was still absent.

To supplement the visual exploration score, which measured visual orientation to the physical environment, visual orientation to the mother and to the stranger were also coded. The only noteworthy findings may be summarized as follows: In episode 2, the baby looked at the toys and other aspects of the physical environment much more

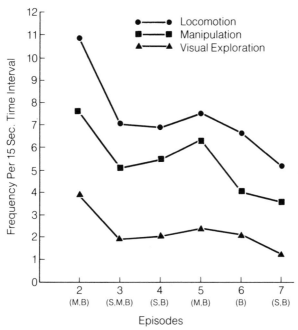

Figure 5.1. Incidence of exploratory behavior.

frequently than at the mother, at whom he glanced only now and then, keeping visual tabs on her; in episode 3, the stranger, the most novel feature of the environment, was looked at more than the toys, and the mother was looked at no more frequently than before.

Crying. Figure 5.2 suggests that the strange situation does not in itself cause alarm or distress, for crying is minimal in episode 2. Crying does not increase significantly in episode 3 ($p = .068$), which suggests that the stranger was not in herself alarming for most Ss, at least not when the mother was also present. The incidence of crying rises in episode 4 with the mother's first departure; it declines upon her return in episode 5, only to increase sharply in episode 6 when she departs a second time, leaving the baby alone. It does not decrease significantly when the stranger returns in episode 7, which suggests that it is the mother's absence rather than mere aloneness that was distressing to most of the babies, and that the greater incidence of crying in episode 6 than in episode 4 is largely due to a cumulative effect.

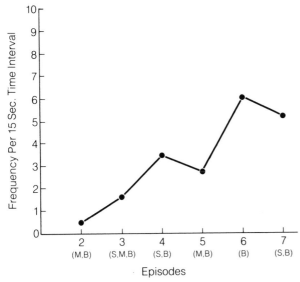

Figure 5.2. Incidence of crying.

Search Behavior during Separation. The mean strength of search behavior was moderate in episode 4 (3.0), significantly stronger in episode 6 (4.6), and moderate again in episode 7 (2.5). Although this might suggest that search behavior is especially activated by being left alone and reduced in the presence of the stranger, this interpretation is not advanced because of the contingencies of the stranger's behavior and her location near the door. Some infants (37 percent) cried minimally if at all in episode 6, and yet searched strongly. Some (20 percent) cried desperately, but searched weakly or not at all. Some (32 percent) both cried and searched. All but four Ss reacted to being left alone with either one or other of these attachment behaviors.

Proximity-Seeking and Contact-Maintaining Behaviors. Figure 5.3 shows that efforts to regain contact, proximity, or interaction with the mother occur only weakly in episodes 2 and 3 but are greatly intensified by brief separation experiences. Contact-maintaining behavior is negligible in episodes 2 and 3, rises in the first reunion episode (5), and rises even more sharply in the second reunion episode (8). In the case of both classes of behavior the increase from episodes 2 through 5 to 8 is highly significant ($p < .001$). Some Ss showed

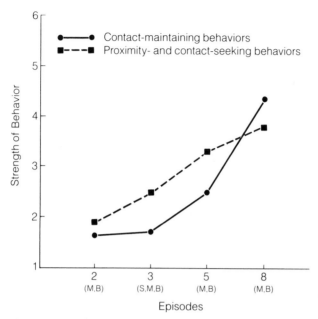

Figure 5.3. Strength of proximity-seeking and contact-maintaining behaviors directed toward the mother.

these behaviors in relation to the stranger also. Thus, for example, a few infants approached the stranger in each of the episodes in which the stranger was present, but substantially fewer than those who approached the mother. Some infants were picked up by th stranger in episodes 4 and 7—in an attempt to comfort them—and some of these did cling to her and/or resist being put down again. Nevertheless proximity-seeking and contact-maintaining behaviors were displayed much less frequently and less strongly to the stranger than to the mother.

Contact-Resisting and Proximity-Avoiding Behaviors. Table 5.1 shows the incidence of contact-resisting and proximity-avoiding behaviors directed to both mother and stranger. Contact-resisting behavior directed toward the mother occurred very rarely in the preseparation episodes because the mother had been instructed not to intervene except in response to the baby's demands, and therefore episodes 2 and 3 are omitted from the table. In the reunion episodes, some *Ss* resisted contact with the mother, but many did not. Therefore table 5.1 shows the incidence of this behavior rather than its mean strength.

About one-third of the sample showed contact-resisting behavior to the mother in episode 5, at least to some degree, and about one-half showed it in episode 8. All but one infant who scored relatively high (4 or higher) in contact-resisting behavior received a comparably high score on contact-maintaining behavior. Thus, at least when directed to the mother, contact-resisting behavior seems to represent classic ambivalence—wanting to be held, wanting to be close, and at the same time angrily resisting contact.

Contact and interaction with the stranger were also resisted but somewhat less frequently than with the mother. Six Ss showed fairly strong contact- or interaction-resisting behavior (scores of 4 or higher) with both stranger in episode 7 and with mother in episode 8, but, for the most part, babies who tended to resist the mother did not resist the stranger and vice versa.

Proximity- and interaction-avoiding behavior did not occur in relation to the mother in the preseparation episodes, for the mother's nonparticipant role made no claim on the baby's attention. But, as shown in table 5.1, it occurred to some degree in about half the sample in each of the reunion episodes, 5 and 8. About one-third of the sample avoided the stranger at some time in episode 3—ignoring her, avoiding meeting her eyes, or moving further away from her. The incidence of

Table 5.1. Incidence of Contact-Resisting and Proximity-Avoiding Behavior to Mother and Stranger

Strength of Behavior	Behavior to Mother		Behavior to Stranger		
	Episode 5	Episode 8	Episode 3	Episode 4	Episode 7
Resist Contact					
6–7	4	6	0	6	7
4–5	5	8	5	3	12
2–3	9	13	2	3	3
1	38	29	49	44	34
Avoid Proximity					
6–7	7	5	4	1	1
4–5	17	13	7	3	6
2–3	3	7	7	1	2
1	29	31	38	51	45

these behaviors declined in episode 4, and even in episode 7 remained less than in episode 3. About half the sample avoided neither mother nor stranger, but those who showed this behavior in any strength (score of 4 or over) to one did not show it to the other.

DISCUSSION

These findings illustrate the complex interaction between attachment behavior, response to novel or unfamiliar stimulus objects and situations, and responses to separation from the attachment object and to subsequent reunion. First, let us consider response to novelty. It is now commonly accepted that novelty may elicit either fear and avoidance or approach and exploration, depending both on the degree of novelty and upon circumstances. One of the conditions which facilitates approach and exploration of the novel is the presence, in reasonable but not necessarily close proximity, of the mother—the object of attachment. The infants of the present sample showed little alarm in the preseparation episodes of the strange situation. Their attachment behavior was not activated; they tended not to cling to the mother or even to approach her. They used her as a secure base from which to explore the strange situation. This finding is not new. Similar observations have been reported by Arsenian (1943), Cox and Campbell (1968), Ainsworth and Wittig (1969), and Rheingold (1969) for human subjects, and by Harlow (1961) for rhesus macaque infants. The presence of the mother can tip the balance in favor of exploring the novel rather than avoiding it or withdrawing from it.

Absence of the mother tends to tip the balance in the opposite direction with a substantial heigtening of attachment behavior and concomitant lessening of exploration. During the mother's absence, proximity-promoting behaviors (crying and search) are evident. The mother's return in the reunion episodes did not serve to redress the balance to its previous level. Attachment behaviors—proximity- and contact-seeking and contact-maintaining behaviors—remained heightened. Crying did not immediately subside in many cases and, despite the mother's attempts to evoke a renewed interest in exploring the properties of the toys, exploration remained depressed below its initial level. . . .

Let us turn from attachment behavior to consider those behaviors that work against contact- and proximity-seeking, namely, contact-resisting and proximity- and interaction-avoiding behaviors. Contact-resisting behavior, as directed toward the mother, usually occurred in conjunction with contact-seeking behavior, and hence, as suggested earlier, implies an ambivalent response. Ambivalent or rejecting and angry responses are reported as common in young children returning home after brief separations (e.g., Heinicke and Westheimer 1965). Separation heightens aggressive behavior of this kind as well as attachment behavior, and predisposes the child toward angry outbursts upon minimal provocation. Spencer-Booth and Hinde (1966) report similar increase of aggression in monkeys: Unusually intense tantrums occur in response to any discouragement of contact-seeking behavior during the period of reunion after separation. Some of our strange-situation *S*s showed contact-resisting behavior toward the stranger. Although in some cases this may indicate fear of the strange person, it seems likely that in some, perhaps most, it is a manifestation of aggression evoked by the mother's departure.

Proximity-avoiding behavior, on the other hand, seems likely to stem from different sources in the case of the stranger than in the case of the mother, even though the overt behavior seems the same in both cases. Ignoring the stranger and looking, turning, or moving away from her probably imply an avoidance of the unfamiliar and fear-evoking person. This is suggested by the fact that these responses are more frequent (as directed toward the stranger) in episode 3, when the stranger has first appeared, than in later episodes. Similar avoidance of the mother cannot be due to unfamiliarity and seems unlikely to be caused by fear. Such behavior occurs in the reunion episodes and is more frequent than avoidance of the stranger.

PROPOSITIONS FOR A COMPREHENSIVE
CONCEPT OF ATTACHMENT

The following propositions are suggested as essential to a comprehensive concept of attachment. They are based on an ethological-evolutionary point of view and have been formulated on the basis of reports of a broad range of investigations, including naturalistic studies of mother-infant interaction, and studies of mother-child separation and reunion in both human and nonhuman primates, as well as the illustrative strange-situation study reported here.

1. Attachment is not coincident with attachment behavior. Attachment behavior may be heightened or diminished by conditions—environmental and intraorganismic—which may be specified empirically. Despite situationally determined waxing and waning of attachment behavior, the individual is nevertheless predisposed intermittently to seek proximity to the object of attachment. It is this predisposition—which may be conceived as having an inner, structural basis—that is the attachment. Its manifestations are accessible to observation over time; a short time-sample may, however, be misleading.

2. Attachment behavior is heightened in situations perceived as threatening, whether it is an external danger or an actual or impending separation from the attachment object that constitutes the threat.

3. When strongly activated, attachment behavior is incompatible with exploratory behavior. On the other hand, the state of being attached, together with the presence of the attachment object, may support and facilitate exploratory behaviors. Provided that there is no threat of separation, the infant is likely to be able to use his mother as a secure base from which to explore, manifesting no alarm in even a strange situation as long as she is present. Under these circumstances the relative absence of attachment behavior—of proximity-promoting behavior—cannot be considered an index of a weak attachment.

4. Although attachment behavior may diminish or even disappear in the course of a prolonged absence from the object of attachment, the attachment is not necessarily diminished; attachment behavior is likely to reemerge in full or heightened strength upon reunion, with or without delay.

5. Although individual differences have not been stressed in this discussion, the incidence of ambivalent (contact-resisting) and probably defensive (proximity-avoiding) patterns of behavior in the reunion episodes of the strange situation are a reflection of the fact that attachment relations are qualitatively different from one attached pair to another. These qualitative differences, together with the sensitivity of attachment behavior to situational determinants, make it very difficult to assess the strength or intensity of an attachment. It is suggested that, in the present state of our knowledge, it is wiser to

explore qualitative differences, and their correlates and antecedents, than to attempt premature quantifications of strength of attachment.

REFERENCES

Ainsworth, M. D. 1964. Patterns of attachment behavior shown by the infant in interaction with his mother. *Merrill-Palmer Quarterly*, 10:51–58.

—— 1967. *Infancy in Uganda: Infant Care and the Growth of Love*. Baltimore: Johns Hopkins University Press.

—— 1969. Object relations, dependency and attachment: a theoretical review of the infant-mother relationship. *Child Development*, 40:369–1025.

Ainsworth, M. D. S. and B. A. Wittig. 1969. Attachment and exploratory behavior of one-year-olds in a strange situation. In B. M. Foss, ed., *Determinants of Infant Behavior*, 4:111–136. London: Methuen.

Arsenian, J. M. 1943. Young children in an insecure situation. *Journal of Abnormal and Social Psychology* 38:225–249.

Bell, S. M. 1970. The development of the concept of the object as related to infant-mother attachment. *Child Development* 41:291–311.

Bowlby, J. 1958. The nature of the child's tie to his mother. *International Journal of Psychoanalysis* 39:350–373.

—— 1969. *Attachment and Loss*. Vol. 1. *Attachment*. London: Hogarth; New York: Basic Books.

Cox, F. N. and D. Campbell. 1968. Young children in a new situation with and without their mothers. *Child Development* 39:123–131.

Hamburg, D. A. 1968. Evolution of emotional responses: evidence from recent research on non-human primates. In J. Masserman, ed., *Science and Psychoanalysis*, 12:39–52. New York: Grune & Stratton.

Harlow, H. F. 1961. The development of affectional patterns in infant monkeys. In B. M. Foss, ed., *Determinants of Infant Behavior*, pp. 75–97. London: Methuen.

Heinicke, C. M. and I. Westheimer. 1965. *Brief Separations*. New York: International Universities Press.

Rheingold, J. L. 1969. The effect of a strange environment on the behavior of infants. In B. M. Foss, ed., *Determinants of Infant Behavior*, 4:137–166. London: Methuen.

Spencer-Booth, Y. and R. A. Hinde. 1966. The effects of separating rhesus monkey infants from their mothers for six days. *Journal of Child Psychology and Psychiatry* 7:179–198.

Six

Growth of Social Play with Peers during the Second Year of Life

Carol O. Eckerman, Judith L. Whatley, and Stuart L. Kutz

We typically do not think of infants as especially social beings, nor do we envision them as having the interest or ability to "play" with other infants. However, as Eckerman, Whatley, and Kutz demonstrate in the following article, this conception may be somewhat misguided. During the second year of life, infants demonstrate increasing interest in play and activities with age-mates. Infants are apparently more sociable at an early age than was previously believed.

IN THE PRESENT STUDY, pairs of normal, home-reared children were brought together in a controlled play setting which included their mothers. The children, similar in age and unfamiliar to one another, were left free to interact with several new toys, their mothers, or one another. The goals were (a) to describe the extent and forms of interaction that the children freely engaged in with one another, (b) to assess changes over the second year of life in their interactions, and (c) to compare their behavior wth one another to that with their mothers and with novel inanimate objects. Such an examination of interactions among young peers is a prerequisite for reasoning about the role of peers in normal human social development.

72

METHOD

Subjects

The subjects were 60 normal, home-reared children, drawn on the basis of age from the population of white infants born and residing in Durham, North Carolina, an industrial city of moderate size. Over 90 percent of the mothers contacted by telephone agreed to participate.

The subjects were divided equally into three age groups—10.0 to 12.0, 16.0 to 18.0, and 22.0 to 24.0 months of age—and paired within each group on the basis of age alone.

Subjects came from homes characterized by above-average levels of education, the equivalent of two or three years of college for the mothers and one or two years of post-college training for the fathers; nevertheless, the range in education was considerable. About half of the children had a sibling, and all but seven spent some time with peers outside the family. Only four children, or 7 percent of the total sample, received 50 percent or more of their daytime care outside the family.

Study Setting

The study took place in a room of moderate size (2.8 × 2.9 m), unfurnished except for a few animal pictures on the walls beyond the subjects' reach and several toys on the floor. The toys were a pulltoy with marbles enclosed in a clear plastic ball, a large plastic dump truck, and three 9-cm-square vinyl cubes decorated with pictures and letters; each toy was present in duplicate. Cushions on the floor in opposite corners along the room's length marked the mothers' positions; the toys were spaced along the wall opposite the mothers' positions. A one-way window behind the toys provided visual access; a microphone in the center of the ceiling, auditory access.

Procedure

Each subject and his mother were escorted to a reception room where they met the other mother and child and the female experimenter. For approximately five minutes the children were left free to sit on their mothers' laps or to explore the room and the few toys it contained while the experimenter instructed the mothers in their role. They were asked to talk naturally with one another, allowing their children to do as they wished; they could respond with a smile or a word or two to the children's social overtures, but they were not to initiate interaction with them or direct their activities unless intervention was necessary to prevent physical harm.

The mothers carried the subjects into the study room, placed them on the floor before the toys, and sat at their positions on the floor. The experimenter then left the room, closing the door behind her, and the 20-minute session began. At the end of the session, the experimenter obtained from the mothers information about the family and the childrens' prior exposure to peers.

Response Measures

An observer behind the one-way window systematically sampled each child's behavior. He focused upon one child at a time and alternated 15-second periods of observation with 15-second periods of recording. Every four observation periods, or 2 minutes, the focus shifted from one child to the other. The resulting record thus was based upon 40 observations, 20 of each child. For each observation, the observer recorded whether or not each of 23 behaviors occurred; the frequency of the behavior within the 15-second period was not recorded.

RESULTS

Reactions to the Novel Play Setting

The subjects of all ages interacted with the toys and their peers, and they contacted their mothers little (see figure 6.1). A multivariate analysis of variance on the three measures indicated a reliable change in behavior with age, $F(6, 50) = 3.42$, $p < .01$. Both the frequency of interactions with the peer and contact with the toy increased reliably with age, $F(2, 27) = 5.90$, $p < .01$; $F(2, 27) = 8.38$, $p < .001$; in contrast, contact with the mother decreased, $F(2, 27) = 7.73$, $p < .01$.

Fussing and crying occurred infrequently, during an average

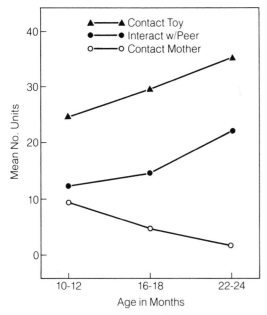

Figure 6.1. General reaction to the novel play setting.

of only one of the 40 periods at each age; and at each age, smiling or laughing occurred more frequently (M = 2.6, 4.7, and 5.1 periods for the 10–12-, 16–18-, and 22–24-month groups, respectively). Some fussing or crying occurred in 14 pairs; smiling or laughing, in 22 pairs.

Nature of the Interactions with the Peer

The subjects attended to the peer (i.e., one or more peer-related behaviors occurred) in over 60 percent of the observations at each age. A multivariate analysis of variance of the five main categories of peer-related behaviors showed a reliable change with age, $F (10, 46)$ = 2.06, p < .05. At all ages a prominent activity was watching the peer. Distant social responses were more frequent at each age than physical contact, but neither changed reliably with age. Vocalizing and smiling were the most frequent of the distant social responses; touching, the most frequent form of contact, especially at the youngest age. The behaviors related to the peer that changed with age were those that involved the play materials. The frequency of both contact with the same play material and direct involvement in the peer's play increased reliably, $F(2, 27) = 4.77$, $p < .02$; $F(2, 27) = 8.42$, $p < .001$, respectively.

Of the behaviors composing direct involvement in the peer's play, four showed orderly increases in frequency with age—imitation of the peer's activity, taking a toy from the peer, struggling over a toy, and coordinating activities with the toys. Five of the behaviors composing direct involvement in the peer's play—offering a toy, accepting a toy, taking a toy, taking over a toy, and struggling over a toy—concern the exchange or attempted exchange of play material between the two children. These exchange activities considered together occurred during an average of 1.9, 3.2, and 5.6 of the 40 intervals at the three ages and thus accounted for the greatest proportion of the activities composing direct involvement in play. Imitation by the focus child was the next most frequent activity across the ages, followed by co-ordinated play. Note, however, that imitation was recorded only when the focus child imitated the peer and not when the peer imitated him; the best estimate, then, of the frequency of imitation by either child is twice the frequencies tabulated, or 1.8, 2.6, and 4.0 periods at the increasing ages. Taking a toy and struggling—negative social responses according to Maudry and Nekula (1939)—together accounted for far less than half of the direct involvement in the peer's play at all ages.

Most of the peer-related behaviors were divided for purposes of comparison into Maudry and Nekula's (1939) positive and negative social reactions. Smile, laugh, vocalize, gesture, touch, show a toy, imitate, offer a toy, accept a toy, and coordinate toy activities comprise the positive reactions. Watching is excluded here, although included by Maudry and Nekula, since its high frequency at all ages would obscure comparison of the remaining behaviors. Fuss, cry, strike, take a toy, take over a toy, and struggle comprise the negative reactions. The ratio, in mean frequency, of the so-called negative reactions to positive reactions was 1.0 to 7.6, 2.2 to 7.8, and 5.2 to 12.6 for the 10–12-, 16–18-, and 22–24-month groups, respectively. At all ages, then, positive responses far outweighed negative ones, even when watching the peer was disregarded; and both positive and negative responses increased with age.

Growth of Social Play

Figure 6.2 contrasts the subjects' social play with solitary play. Social play includes the prior categories of same play material and

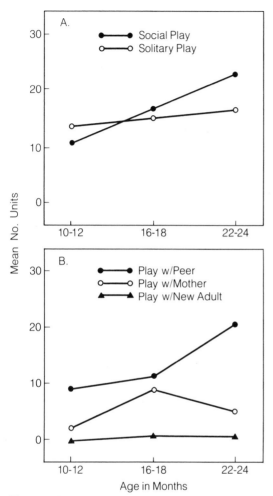

Figure 6.2. The development of social play. (In social play, the child involves others—peer, mother, or new adult—in his activities with nonsocial objects.)

direct involvement in play with the peer, as well as play with the mothers. A multivariate analysis of variance of solitary play and social play showed a reliable change in behavior with age. $F(4, 52) = 3.78$, $p < .01$. Solitary play occurred in slightly more than one-third of the periods at all ages; social play, in contrast, increased reliably with age, $F(2, 27) = 5.42$, $p < .01$, until by 2 years it occurred in 60 percent of the periods. The increase in social play resulted from the reliable increase in play with the peer (see figure 6.2B), $F(2, 27) =$

10.86, $p < .001$. Play with the mother reached a peak at 16 to 18 months, $F(2, 27) = 3.28$, $p = .05$; play with the new adult, that is, the other mother, was rare. By 2 years of age, then, most of the children's activities with toys also involved a person, and in this novel play setting, that person was more often the peer than the mother.

DISCUSSION

The results showed that during the second year of life, children in a novel play setting turned less to their mothers and more and more to inanimate objects and their peers. Further, the children increasingly integrated their activities with toys and peers until by two years of age, social play predominated. They touched the same objects as their peers, imitated their actions, exchanged toys, and coordinated their play to perform a common task or elaborate a social game.

The integration of activities with toys and people during the second year of life highlights a central aspect of development. The responses we call social do not develop in a context devoid of inanimate objects; similarly, those responses toward inanimate objects that we label exploration or play often occur in a social context. The mother, or attachment object, provides the setting for the infant's exploration of his inanimate world (e.g., Ainsworth and Wittig 1969; Rheingold 1969). Social objects, adults or children, also alter the stimulation that toys offer; they put toys in motion, make them do new things, and place them with other objects in new combinations. A toy on the ground does not equal a toy in the hand of a peer. On the other hand, nonsocial objects are vehicles for many forms of social interaction—for giving and taking, imitating, cooperating, or instructing. Smiling and vocalizing to persons or contacting them may be prominent early social behaviors, but other forms of social interaction require the child's integration of activities with things and people.

Questions remain about the determinants and function of the various forms of early interaction between peers. Are watching a peer's activities, looking, smiling, and vocalizing at him, and offering him toys functionally similar to other behaviors we label exploratory? Is the similarity perceived in matching one's activities to another's reinforcing? Do toy exchanges function to promote imitation and same play? Questions such as these await detailed study of the context in which the peer-related behaviors occur. Still, the present results establish several forms of peer interaction whose origins should be sought in infancy rather than the preschool years, and the behavioral descriptions of these interactions provide the starting point for studies of origins and function.

The children of all ages behaved toward one another not as children do with inanimate objects, but rather more as they do with familiar adults. The physical contact and manipulation characteristic of early activity with toys were minimal with the peer; rather the smiles, vocalizations, and offers of toys to the peer correspond to behaviors seen with mothers (e.g., Rheingold 1973). A comparison, however, between play with the mother and direct play with the peer (the most nearly comparable categories of play) yields suggestive differences in the developmental course of these activities. At 10 to 12 months both were infrequent; by 16 to 18 months both were more frequent, but play with the mother exceeded that with the peer; and by 22 to 24 months, play with the mother had declined in frequency, while direct play with the peer increased markedly. Both the correspondence in forms of behavior toward the mother and the peer and the initial greater frequency of these behaviors toward the mother suggest that young children generalize to peers responses developed through interaction with familiar adults.

By two years of age, however, the children of the present study interacted more with the peer than with their mothers. The restrictive instructions to the mother and the short duration of the play period may have contributed to this preference, but the fact remains that with increasing age primate young turn progressively more to their peers in play (e.g., Harlow 1969; Heathers 1955). The question of what peers offer in contrast to familiar adults is an important one for the fuller understanding of early social behavior. Peers and adults seem to differ in their persistence and affective involvement in children's play and the contrast shifts during the child's development. Mothers and fathers initially invest great persistence, affect, and imagination in attempts to engage their young infants in play, but later in development peers often seem more ready than parents to follow through on a child's playful overtures. Still, peers have two other characteristics that warrant study: Their actions and reactions may be more novel than those of parents, and the activities of peers may be more easily duplicated than those of adults, who only with difficulty behave as a child.

The present study, then, has demonstrated that during the second year of life home-reared children freely engage in a variety of social interactions with unfamiliar peers, interactions that resemble those with familiar adults more than those with inanimate objects. They smile, laugh, vocalize, and gesture to one another, show and offer toys, imitate each other, struggle, and engage in reciprocal play. Thus, interactions with peers and with familiar adults appear closely intertwined in development. Yet as more and more of the children's play involved people, the social partner was more often the peer than the mother. This choice of the peer makes still more tenable the speculation that peers—even infant peers—make their own contribution to early human sociability.

REFERENCES

Ainsworth, M. D. S. and B. A. Wittig. 1969. Attachment and exploratory behaviour of one-year-olds in a strange situation. In B. M. Foss, ed., *Determinants of Infant Behaviour*, vol. 4. London: Methuen.

Bridges, K. M. B. 1933. A study of social development in early infancy. *Child Development* 4:36–49.

Bühler, C. 1930. *The First Year of Life*. New York: John Day.

Harlow, H. F. 1969. Age-mate or peer affectional system. In D. S. Lehrman, R. A. Hinde, and E. Shaw, eds., *Advances in the Study of Behavior*, vol. 2. New York: Academic Press.

Heathers, G. 1955. Emotional dependence and independence in nursery school play. *Journal of Genetic Psychology* 87:37–57.

Hinde, R. A. 1971. Development of social behavior. In A. M. Schrier and F. Stollnitz, eds. *Behavior of Nonhuman Primates: Modern Research Trends*, vol. 3. New York: Academic Press.

Maudry, M. and M. Nekula. 1939. Social relations between children of the same age during the first two years of life. *Journal of Genetic Psychology* 54:193–215.

Rheingold, J. L. 1969. The effect of a strange environment on the behaviour of infants. In B. M. Foss, ed., *Determinants of Infant Behavior*, vol. 4. London: Methuen.

—— 1973. Independent behavior of the human infant. In *Minnesota Symposia on Child Psychology*, vol. 7. Minneapolis: University of Minnesota Press.

Rheingold, H. L. and C. O. Eckerman. 1973. Fear of the stranger: a critical examination. In H. W. Reese, ed., *Advances in Child Development and Behavior*, vol. 8. New York: Academic Press.

Part II

Childhood
Mastery and Competencies

Seven

The Role of Play in Cognitive Development
Brian Sutton-Smith

In this selection, Brian Sutton-Smith discusses an influence on the child's cognitive development that is often overlooked—play. What is the function of play in terms of the child's developing sense of logic and rules, growing understanding of social roles and social relationships, and acquisition of information about the environment and the culture? Sutton-Smith suggests that in many important ways, play is much more than mere "fun and games."

THE FUNCTION OF PLAY

DESPITE THE FACT that a great deal has been written about play, there is actually very little research on the subject matter of the play function itself. That is, very little is known about what play accomplishes for human or animal organisms. This neglect of play's "function" seems to have occurred historically because of the key role of "work" in industrial civilization and the concomitant derogation of the importance of recreation and leisure (de Grazia 1962). In addition, and perhaps for similar historical reasons, explanations of behavior both in biological and psychological thinking have been serious and utilitarian in nature; that is, an activity has not been thought to be explained unless its direct value for the organism's survival could be indicated. For this reason, play, which on the surface at least is not a very useful activity, has been interpreted most often as an illustration of the working of other "useful" functions, rather than a peculiar function in its own right. It has been said that in play the child "reduces tensions," "masters anxiety," "generalizes responses," or manifests

83

a polarity of "pure assimilation." In each of these cases the explanation of play has been subsumed to the workings of theoretical concepts which could just as well be illustrated without reference to play.

In consequence, the research literature on play is mainly about variables that are not necessarily central to an understanding of play itself. For example, levels of social development (e.g., solitary, parallel, and the like) are said to be illustrated in the play of preschoolers; more severely punished children are said to express more aggression in their doll play; children are said to prefer to go on with play activities that have been interrupted; play and game preferences are used as evidence of sex-role identification, anxiety, intellectual level, race, environment, need achievement, levels of aspiration, and sociometric status (Marshall 1931; Hurlock 1934; Levin and Wardwell 1962; Sutton-Smith, Rosenberg, and Morgan 1963). In all these cases the character of the play is treated as epiphenomenal to the other more fundamental variables with which the researcher is concerned. In the literature of the past few years, however, there has developed a changing attitude toward the functional significance of play and games, and it is to this literature that the present article will be devoted. . . .

Having said that play's function is beginning to be explored in its own right, however, and that this is in turn an outcome of current trends in social science, one has then to admit that there is as yet no generally accepted definition of what play really is or what it does. . . .

In a great deal of current research and as a part of play's rehabilitation as serious subject matter, play has generally been identified with exploratory behavior. Both exploratory behavior and play have been described as self-motivated activities whose rewards lie in the gratifications that they bring directly to the participants (Berlyne 1960). One typical finding from this work, emerging from many animal as well as some human studies, is that novel properties in the ecology (blocks, puzzles, colors, and games) increase the response levels of the subjects exposed to those properties. As subjects cease to be able to do new things with objects, however, their response to them decreases. Berlyne has indicated that other properties of objects which have similar effects are their complexity, their surprisingness, their uncertainty, and their capacity to induce conflict. It has also been found that the greatest increases in response level are recorded for those objects with which the subjects can do most things, that is, which can be handled, moved, seen, touched, and so forth. Further, exploratory and play behavior, like other response systems, is susceptible to increase or diminution in response level as a result of appropriate parental reinforcements (Aldrich 1965; Marshall 1966). Finally, exploratory and play behavior in child subjects correlates highly with information seeking in general (Maw and Maw 1965).

Unfortunately, because play and exploration are categorized together in most of these studies, it is not possible to state what proportion of the increased responsivity is due to one or the other. But attempts have been made to distinguish between the two. On the basis of his observations of infants, for example, Piaget cast them into a temporal relation, with exploration preceding play:

We find, indeed, though naturally without being able to trace any definite boundary, that the child, after showing by his seriousness that he is making a real effort at accommodation, later produces these behaviors merely for pleasure, accompanied by smiles and even laughter, and with the exception of results, characteristic of the circular reactions through which the child learns. (1951:90)

Welker, on the other hand, sees the difference mainly in terms of a passivity-activity dimension. Thus, he says exploration "consists of cautiously and gradually exposing the receptors . . . to portions of the environment. The goals or incentives consist of sensory stimulation, and novel stimuli in any modality are especially important. *Play* consists of a wide variety of vigorous and spirited activities: those that move the organism or its parts through space such as running, jumping, rolling . . . and vigorous manipulation of body parts or objects in a variety of ways" (1961:176). In Welker's account the major emphasis is upon the novel variation of the subject's own responses, more or less irrespective of the variation of the stimulus qualities of the objects with which he is engaged. It is very clear in Piaget's observations and statements that these distinctions are difficult to make because exploration and play are both polarities within self-motivated activity with the child often changing rapidly from one to the other, so that it is difficult to classify an activity as one or the other. It is only in older subjects, in fact, who themselves categorize their activities as one or the other, as hobbies or games, that we become reasonably certain of the difference. Still, for the purposes of this paper, we will provisionally suggest that play, while like exploratory behavior in being intrinsically motivated, is different from the latter in its greater emphasis upon the novel variation of responses according to internal criteria; play is an activity accompanied by the traditional and often-mentioned affective accompaniments of "playfulness," "fun," and "the enjoyment of the activity for its own sake."

PLAY AND COGNITION

Given this conception of play, we are in a position to ask what cognitive difference such variation seeking can make. In classical psychoanalytic and Piagetian theory, the play of the child is said to have a mainly compensatory function. For the analysts, play has little significance for intellectual growth except as it helps to reduce the amount of tension that might be impeding intellectual activity somewhere else. For Piaget, play permits the child to make an intellectual response in fantasy when he cannot make one in reality, and this protects his sense of autonomy. In addition, however, it helps to consolidate learnings acquired elsewhere and prevents them from dropping into disuse. These two viewpoints may be contrasted with others in which the play itself is given a much more active cognitive function in the development of thought. Psychoanalyst Erik Erikson suggests that the young child's play is analogous to the planning of an adult. Several generations of sociologists likewise have seen play as providing model situations in which the child rehearses roles he will later occupy seriously somewhere else. While most of these sociologists emphasize the social value of the play, some also stress cognitive implications. For example, George H. Mead stated that children develop social *understanding* through having to take the role of the other into account in their own actions. That is, the child cannot hide very successfully in Hide-and Seek unless he has also taken into account what happens when someone seeks (Mead 1934; Goffman 1961).

But these are general theoretical viewpoints whereas our interest here is in research investigations of play as form of cognitive variation seeking. A useful lead is provided by the work of Lieberman (1965). She was interested in relations between children's playfulness and their creativity. Her subjects were 93 kindergarten children from middle-class homes attending five kindergarten classes in three New York schools. The children were rated on playfulness scales which included the following characteristics:

1. *How often does the child engage in spontaneous physical movement and activity during play?* This behavior would include skipping, hopping, jumping, and other rhythmic movements of the whole body or parts of the body like arms, legs, or head, which could be judged as a fairly clear indication of exuberance.

2. *How often does the child show joy in or during his play activities?* This may be judged by facial expression such as smiling, by verbal expressions such as saying "I like this" or "This is fun" or by more indirect vocalizing such as singing as an accompaniment of the activity, e.g., "choo, choo, train go along." Other behavioral indicators would be repetition of activity or resumption of activity with clear evidence of enjoyment.

3. *How often does the child show a sense of humor during play?* By "sense of humor" is meant rhyming and gentle teasing ("glint-in-the-eye" behavior), as well as an ability to see a situation as funny as it pertains to himself or others.

4. *While playing, how often does the child show flexibility in his interaction with the surrounding group structure?* This may be judged by the child joining different groups at any one play period and becoming part of them and their play activity, and by being able to move in and out of these groups by his own choice or by suggestion from the group members without aggressive intent on their part.

A factor analysis of the results led Lieberman to conclude that these scales tapped a single factor of playfulness in these children. But the finding to which we wish to call attention in the present case is the significant relation which was found between playfulness and ability on several creative tasks. That is, children who were rated as more playful were also better at such tasks as: suggesting novel ideas about how a toy dog and a toy doll could be changed to make them more fun to play with; giving novel plot titles for two illustrated stories that were read and shown to the children; and giving novel lists of animals, things to eat, and toys. Unfortunately, the problem with Lieberman's work, as well as with much other work involving creativity measures, is that intelligence loads more heavily on the separate variables of playfulness and creativity than these latter variables relate to each other. Consequently, we cannot be sure whether the findings reflect a distinctive relation between playfulness and creativity or whether these variables are two separate manifestations of intelligence as measured by conventional intelligence tests.

And yet it seems to make sense that the variations in response which constitute playful exercise should be similar to the required variations in response on creativity tests. In other words, these two variables appear to be structurally similar. Our confidence that this may indeed be the case is bolstered by some work of Wallach and Kogan (1965) who found that if they gave their creativity tests in a situation in which the subjects were free from usual test pressures, they did indeed obtain creativity scores which were in the main statistically distinct from conventional intelligence test scores. Their conditions for producing these results were individual testing, a complete freedom from time pressures, and a *game-like approach* to the task. The experimenters were introduced to the subject as visitors interested in children's games, and for several weeks prior to testing, spent time with the children in an endeavor to heighten this impression. From this work, Wallach and Kogan concluded that creativity is indeed something different from conventional intelligence and that its manifestation is facilitated in a playful atmosphere. In consequence it may be concluded that if playfulness and creativity co-vary as Lieberman discovered, it is not a function of their separate relations to intelligence.

PLAY AND NOVEL REPERTOIRES

What then is the functional relation between the two? While there are various possibilities, only one will be presented here as the concern is more with research than it is with theory. The viewpoint taken is that when a child plays with particular objects, varying his responses with them playfully, he increases the range of his associations for those particular objects. In addition, he discovers many more uses for those objects than he would otherwise. Some of these usages may be unique to himself and many will be "imaginative," "fantastic," "absurd," and perhaps "serendipitous." Presumably, almost anything in the child's repertoire of responses or cognitions can thus be combined with anything else for a novel result, though we would naturally expect recent and intense experiences to play a salient role. While it is probable that most of this associative and combinatorial activity is of no utility except as a self-expressive, self-rewarding exercise, it is also probable that this activity increases the child's repertoire of responses and cognitions so that if he is asked a "creativity" question involving similar objects and associations, he is more likely to be able to make a unique (that is, creative) response. This is to say that play increases the child's repertoire of responses, an increase which has potential value (though no inevitable utility) for subsequent adaptive responses.

In order to test this relation, the writer hypothesized that children would show a greater repertoire of responses for those toys with which they had played a great deal than for those with which they had played less. More specifically, it was hypothesized that both boys and girls would have a greater repertoire of responses with objects for their own sex than for opposite sex objects. In order to control for differences in familiarity, like and opposite sex toys were chosen that were familiar to all subjects. Four toys were selected that had been favorites during the children's year in kindergarten. The girls' toys were dolls and dishes; the boys' were trucks and blocks. It was expected that as they had all known and seen a great deal of all of these toys throughout the year, they would not differ in their familiarity with the toys, as measured by their descriptions of them, but that they would differ in their response variations with these toys as measured by their accounts of the usages to which the toys could be put. Nine boys and nine girls of kindergarten age were individually interviewed, and the investigator played the "blind" game with them. That is, of each toy, he asked, pretending that he was blind: "What is it like?" (description), and "What can you do with it?" (usage). Each child responded to each toy. The interviews were conducted in a leisurely manner, the longest taking 45 minutes and the most usages given for one object being 72 items. The results were that the sexes did not differ from each other in their descriptions of the four objects. Both sexes did differ, however, in the total number of usages given for each toy and the number of unique usages. Boys were able to give more usages and more unique usages for trucks and blocks than they could give for dolls and dishes, although they had not differed between the two sets in their descriptions. Similarly, the girls displayed a larger repertoire for the objects with which they had most often played, dolls and dishes, than for trucks and blocks which had also been in the kindergarten all year, but with which they had not played extensively (Sutton-Smith 1967).

As the number of responses was not related to intelligence, and as the children showed equal familiarity with all objects (as judged by their descriptions), it seemed reasonable to interpret their response to this adaptive situation (asking them questions) as an example of the way in which responses developed in play may be put to adaptive use when there is a demand. This principle may apply to games as well as play. While most of the activities that players exercise in games have an expressive value in and for themselves, occasionally such activities turn out to have adaptive value, as when the subject, a healthy sportsman, is required in an emergency to run for help, or when the baseball pitcher is required to throw a stone at an attacking dog, or when the footballer is required to indulge in physical combat in war, or when the poker player is required to consider the possibility that a business opponent is merely bluffing. In these cases, we need not postulate any very direct causal connection between the sphere of play and the sphere of adaptive behavior, only the general evolutionary requirement that organisms or individuals with wider ranges of expressive characteristics, of which play is but one example, are equipped with larger response repertoires for use in times of adaptive requirement or crisis. . . .

PLAY AND THE REPRESENTATIONAL SET

But there is perhaps an even more essential way in which play might be related to cognition. Beginning with the representational play of two-year-olds, there develops a deliberate adoption of an "as if" attitude toward play objects and events. The child having such an attitude continues to "conserve" imaginative identities throughout the play in spite of contraindicative stimuli. This cognitive competence is observable both in solitary play, social games, and in the children's appreciation of imaginative stories, Yet it is not until five to seven years of age that children can conserve the class identities of such phenomena as number, quantity, space, and the like, despite contraindicative stimuli. Paradoxically the factor which prevents children from conservation of class identities appears to be the very stimulus bondedness which they are able to ignore in their play. The question can be raised, therefore, as to whether the ability to adopt an "as if" or representational set in play has anything to do with the ability to adopt representative categories on a conceptual level. The only available data are correlational in nature, but again they show a correspondence between the status of the play and the status of the cognition. In Sigel's studies of cognitive activity, lower-class children who exhibited an inability to categorize in representational terms were also impoverished in their play, showing a high frequency of motoric activity, minimal role playing, and block play of low elaboration (Sigel and McBane 1966). The evidence suggests the possibility that play may not only increase the repertoire of available responses, but that, where encouraged, it may also heighten the case with which representational sets can be adopted toward diverse materials.

The difficulty with the studies so far cited, however, is that we cannot be sure whether play merely expresses a preexisting cognitive status of the subjects or whether it contributes actively to the character of that status. That is, is the play constitutive of thought or merely expressive of thought? More simply, does the player learn anything by playing?

PLAY AS LEARNING

The view that something is learned by play and games has long been a staple assumption in the "play way" theory of education and has been revived among modern educators under the rubric of game simulation (Bruner 1965; Meier and Duke 1966). Evidence for effects of particular games on particular learnings are few, although where research has been carried out, it seems to be of confirming import. Research with games involving verbal and number cue seems to show that games result in greater improvement than occurs when control groups receive the same training from more orthodox workbook procedures (Humphrey 1965, 1966). Similarly, research with games requiring the exercise of a variety of self-controls seems to indicate social improvement in the players (Gump and Sutton-Smith 1955; Sutton-Smith 1955; Redl 1958; Minuchin, Chamberlain, and Graubard 1966).

As an example of this type of field research, the present investigator used a number game to induce number conservation in young children between the ages of 5.0 and 5.7 years. The game known traditionally as "How many eggs in my bush?" is a guessing game in which the players each hide a number of counters within their fists, and the other player must guess the number obscured. If he guesses correctly, the counters are his. The players take turns and the winner is the player who finishes up with all the counters. Each player begins with about 10 counters. Children in the experimental group showed a significant improvement from a pre- to post-test on number conservation as compared with children in the control group. The game apparently forced the players to pay attention to the cues for number identity or they would lose, be cheated against, be laughed at, and would certainly not win (Sutton-Smith 1967).

Given these demonstrations that learning can result as a consequence of game playing, we are perhaps in a better position to interpret those other studies of games which show that continued involvement in games is correlated with important individual differences in player personality and cognitive style. For example, a series of studies has been carried out with the game of Ticktacktoe (Sutton-Smith, Roberts, et al. 1967). Ticktacktoe is the most widespread elementary game of strategy and is a game in which players compete to see who can get three crosses or circles in a row on a grid-shaped diagram. A series of studies with this game has shown that children who are better players are indeed very different from those who are losers. More important, distinctions have been established between those who tend to win on this game and those who tend to draw. Although these chidren do not differ in intelligence, they do differ in a number of other ways. Boys who are winners are also perceived as "strategists" by their peers on a sociometric instrument. They are better at arithmetic; they persevere at intellectual tasks; they are rapid at making decisions. Boys who are drawers, on the other hand, are less independent, more dependent on parents and teachers for approval, and more conventional in their intellectual aspirations. Girls who are winners are aggressive and tomboyish, whereas girls who are drawers are withdrawing and ladylike. These results support the view that there are functional interrelations between the skills learned in games and other aspects of player personality and cognitive style.

Similarly, cross-cultural work with games seems to show that games are tied in a functionally enculturative manner to the cultures of which they are a part. Thus, games of physical skill have been shown to occur in cultures where there is spear-throwing and hunting.

The older tribal members introduce and sustain these games which have a clear-cut training value.

Games of chance occur in cultures where there is punishment for personal achievement and an emphasis upon reliance on divinatory approaches to decision-making (Roberts and Sutton-Smith 1966); games of strategy occur in cultures where the emphasis is on obedience and diplomacy as required in class and intergroup relations and warfare (Roberts, Sutton-Smith, and Kendon 1963).

Still, all this research, though it implies functional relations between games and culture patterns, and between games and cognitive styles is like the pedagogic research mentioned above. The latter clearly demonstrates that one can gain a pedagogic and cognitive advantage by use of games for training purposes, but the research is weak insofar as it does not allow us to draw conclusions concerning the particular facets of the games that have the observed influence. The multi-dimensional character of play and of games makes it difficult to specify the key variables which are effective in bringing about the cognitive changes. We do not know yet what interaction between player desire to win and attention to the correct cues brings about the demonstrated learning. This is a subject for future research.

In conclusion, the intent of the present account has been to indicate that there is evidence to suggest that play, games, and cognitive development are functionally related. But the relation, it has been stressed, is a loose one. Play, like other expressive characteristics (laughter, humor, and art), does not appear to be adaptive in any strictly utilitarian sense. Rather, it seems possible that such expressive phenomena produce a superabundance of cognitions as well as a readiness for the adoption of an "as if" set, both of which are potentially available if called upon for adaptive or creative requirements. Given the meagerness of research in this area, however, it is necessary to stress that these are conclusions of a most tentative nature.

REFERENCES

Aldrich, N. T. 1965. Children's level of curiosity and natural child-rearing attitudes. Paper presented at Midwestern Psychological Association, Chicago, May.

Berlyne, D. C. 1960. *Conflict, Arousal and Curiosity.* New York: McGraw-Hill.

Bruner, J. S. 1965. Man: a course of study. *Educational Services Inc. Quarterly Report* 3:85–95.

de Grazia, S. 1962. *Of Time, Work, and Leisure.* New York: Twentieth Century Fund.

Goffman, I. 1961. *Encounters.* Indianapolis: Bobbs-Merrill.

Gump, P. V. and B. Sutton-Smith. 1955. The "it" role in children's games. *The Group* 17:3–8.

Humphrey, J. H. 1965. Comparison of the use of active games and language workbook exercises as learning media in the development of language understandings with third grade children. *Percept. Mot. Skills* 21:23–26.

—— 1966. An exploratory study of active games in learning of number concepts by first grade boys and girls. *Percept. Mot. Skills* 23:341, 342.

Hurlock, E. B. 1934. Experimental investigations of childhood play. *Psychological Bulletin* 31:47–66.

Levin H. and E. Wardwell. 1962. The research uses of doll play. *Psychological Bulletin* 59:27–56.

Lieberman, J. N. 1965. Playfulness and divergent thinking: an investigation of their relationship at the kindergarten level. *J. Genet. Psychol* 197:219–224.

Marshall, H. 1931. Children's plays, games, and amusements. In C. Murchison, ed., *Handbook of Child Psychology*, pp. 515–526. Worcester, Mass.: Clark University Press.

Marshall, Helen R. and C. H. Shwu. 1966. Experimental modification of dramatic play. Paper presented at the American Psychological Association, New York, September.

Maw, W. H. and E. W. Maw. 1965. Personal and social variables differentiating children with high and low curiosity. *Cooperative Research Project No. 1511*, pp. 1–181. Wilmington: University of Delaware.

Mead, G. H. 1934. *Mind, Self, and Society.* Chicago: University of Chicago Press.

Meier, R. L. and R. D. Duke. 1966. Game simulation for urban planning. *J. Amer. Institute of Planners* 32:3–18.

Minuchin, Patricia, P. Chamberlain, and P. A. Graubard 1966. A project to teach learning skills to disturbed delinquent children. Paper presented at the 43rd Annual Meeting of the American Orthopsychiatric Association, San Francisco, April.

Piaget, J. 1951. *Play, Dreams, and Imitation in Childhood.* London: Heinmann.

Redl, F. 1958. The impact of game ingredients on children's play behavior. *Fourth Conference on Group Processes*, pp. 33–81. New York: Josiah Macy Grant Foundation.

Roberts, J. J. and B. Sutton-Smith. 1966. Cross cultural correlates of games of chance. *Behav. Sci. Notes* 3:131–144.

Roberts, J. M., B. Sutton-Smith, and A. Kendon. 1963. Strategy in folk-tales and games. *J. Soc. Psychol.* 61:185–199.

Sigel, I. E. and B. McBane 1966. Cognitive competence and level of symbolization among five year old children. Paper read at American Psychological Association, New York, September.

Sutton-Smith, B. 1967. A game of number conservation. Unpublished manuscript, Bowling Green State University.

—— 1959. *The Games of New Zealand Children.* Berkeley: University of California Press.

—— 1967. Novel signifiers in play. Unpublished manuscript, Bowling Green State University.

—— 1955. The psychology of games. *National Education*, pt. 1, pp. 228–229, and pt. 2, pp. 261, 263 (Journal of New Zealand Educational Institute).

Sutton-Smith, B., J. M. Roberts, et al. 1967. Studies in an elementary game of strategy. *Genet. Psychol. Monogr.* 75:3–42.

Sutton-Smith, B., B. G. Rosenberg, and E. Morgan. 1963. The development of sex differences in play choices during preadolescence. *Child Developm* 34:119–126.

Wallach, M. A. and N. Kogan. 1965. *Modes of Thinking in Young Children.* New York: Holt, Rinehart & Winston.

Welker, W. I. 1961. An analysis of exploratory and play behavior in animals. In D. W. Fiske and S. R. Maddi, eds., *Functions of Varied Experience.* Homewood, Ill.: Dorsey.

Eight

A Structural-Developmental Analysis of Levels of Role-taking in Middle Childhood

Robert L. Selman and Diane F. Byrne

Researchers have become increasingly interested over the past decade in the child's development of what is called *social cognition*—**how the individual thinks about others and about relationships with them. One important aspect of social cognition is role-taking, "the ability to put oneself in another's place and view the world through his eyes." In this selection, Selman and Byrne present a developmental perspective on changes in role-taking abilities during childhood and suggest that the child's social cognitive abilities develop along lines parallel to the development of cognitive abilities in general.**

THE PARALLEL STRUCTURAL development of impersonal and interpersonal cognitions has been posited by a number of theorists (Mead 1934; Piaget 1950). Piaget states: "There is a fundamental identity between the interpersonal operations and the intraindividual operations so that they can be isolated only by abstraction from a totality where the biological and social factors of action constantly interact with one another" (1967:129).

In this paper we focus on one aspect of interpersonal cognition—social role-taking—and attempt to define its development according to an ontogenetic sequence of structures similar in form to Piaget's cognitive operations.

Social role-taking has a long-standing tradition as a theoretical concept of basic importance to developmental and social psychology. The theoretical writings of George Herbert Mead (1934) and James

96

Mark Baldwin (1906) support the position that the unique aspect of social cognition and judgment that differentiates human from sub-human functioning is "role-taking," the ability to understand the self and others as subjects, to react to others as like the self, and to react to the self's behavior from the other's point of view. The concept of role-taking also has roots in Piaget's theory of cognitive development. Two of his central concepts relate directly to role-taking: egocentrism, which characterizes preoperational thinking, is the inability to escape from one's own view of the world; decentration, a characteristic of operational thinking, is the ability to consider multiple perspectives or aspects of a situation. While these concepts are applicable to the impersonal domain, they also apply to the interpersonal sphere and point to development in the ability to put onself in another's place and view the world through his eyes.

There have been two recent approaches to the study of social role-taking stemming from the Piagetian point of view that have influenced the present research. Feffer (1959, 1971), and Feffer and Gourevitch (1960) equate social role-taking with the Piagetian concept of social decentering and have developed a projective technique for assessing age-related levels of the child's ability to decenter in the social domain. Feffer has described a series of formal levels of this ability: simple refocusing, characterized by a lack of coordination between perspectives; consistent elaboration, defined as a sequential coordination between perspectives; and change of perspective, at which simultaneous coordination of perspectives is achieved.

A second important attempt to clarify the role-taking concept through systematic empirical investigation is Flavell's (1968) study of the development of children's ability to make inferences about another's perceptual or conceptual perspectives. Flavell has isolated three crucial steps in the development of role-taking ability. The first is self's recognition that other can have cognitions about the self as well as about other external objects. The second discovery that must be made is that the self is not only an object for other, but also a subject. The third achievement is the recognition that both self and other can go on considering each other's view of the other ad infinitum (p. 53).

The structural-developmental approach to role-taking, then, is the derivation of a sequence of developmental age-related and logically related structures or forms that an individual displays in his understanding of other's point of view. The concern is not with content, not with accuracy of perception of other or behavioral choice, but with the form in which conceptions of others emerge.

While Feffer has explored role-taking within the context of a projective story-telling task and Flavell within social problem-solving and communication tasks, in the present study we have focused on role-taking as it is used within the context of moral dilemmas similar to those developed by Kohlberg (1969), but modified to be more appropriate for young children. We constructed a series of four role-taking levels on the basis of Feffer and Falvell's analyses, the previous research of Selman (1971), and developmental principles of differentiation (distinguishing perspectives) and integration (relating perspectives). The levels thus derived were tested on the experimental sample. A brief description of each level follows.

LEVEL 0: EGOCENTRIC ROLE-TAKING

Distinguishing perspectives. This stage is characterized by the child's inability to make a distinction between a personal interpretation of social action (either by self or other) and what he considers the true or correct perspective. Therefore, although the child can differentiate self and other as entities, he does not differentiate their points of view.

Relating Perspectives. Just as the child does not differentiate points of view, he does not relate perspectives.

LEVEL 1: SUBJECTIVE ROLE-TAKING

Distinguishing Perspectives. At level 1 the child sees himself and other as actors with potentially different interpretations of the same social situation, largely determined by the data they have at hand. He realizes that people feel differently or think differently because they are in different situations or have different information.

Relating Perspectives. The child is still unable to maintain his own perspective and simultaneously put himself in the place of others in attempting to judge their actions. Nor can he judge his own actions from their viewpoint. He has yet to see reciprocity between perspectives, to consider that his view of other is influenced by his understanding of other's view of him (level 2). He understands the subjectivity of persons but does not understand that persons consider each other as subjects rather than only as social objects.

LEVEL 2: SELF-REFLECTIVE ROLE-TAKING

Distinguishing Perspectives. The child is now aware that people think or feel differently because each person has his own uniquely ordered set of values or purposes.

Relating Perspectives. A major development at level 2 is the ability to reflect on the self's behavior and motivation as seen from outside the self, from the other's point of view. The child recognizes that the other, too, can put himself in the child's shoes, so the child is able to anticipate other's reactions to his own motives or purposes. However, these reflections do not occur simultaneously or mutually. They only occur sequentially. The child cannot "get outside" the two-person situation and view it from a third-person perspective.

LEVEL 3: MUTUAL ROLE-TAKING

Distinguishing Perspectives. The child can now differentiate the self's perspective from the generalized perspective, the point of view taken by some average member of a group. In a dyadic situation he distinguishes each party's point of view from that of a third person. He can conceive of the concept of "spectator" and maintain a disinterested point of view.

Relating Perspectives. The child at level 3 discovers that both self and other can consider each party's point of view simultaneously and mutually. Each can put himself in the other's place and view himself from that vantage point before deciding how to react. In addition, each can consider a situation from the perspective of a third party who can also assume each individual's point of view and consider the relationships involved.

Method

Subjects. The Ss were forty middle-class children, ten each at ages 4, 6, 8, and 10. In each age group, there were five males and five females.

Task. The Ss were given two open-ended dilemmas, each presented in the form of a filmstrip. A sample dilemma with some standard probe questions is described below.

Holly is an 8-year-old girl who likes to climb trees. She is the best tree climber in the neighborhood. One day while climbing down from a tall tree she falls off the bottom branch but does not hurt herself. Her father sees her fall. he is upset and asks her to promise not to climb trees anymore. Holly promises.

Later, that day, Holly and her friends meet Sean. Sean's kitten is caught up in a tree and cannot get down. Something has to be done right away or the kitten may fall. Holly is the only one who climbs trees well enough to reach the kitten and get it down, but she remembers her promise to her father.

The average interview time for each dilemma was 20–25 min. At the end of the filmstrip, each S was asked to retell the story before questioning so that the E could be assured that any difficulties were not simply due to faulty memory. Few Ss had difficulty in retelling the story. If difficulty was encountered, the dilemma was repeated. Interviews were taped and transcribed for scoring purposes. Standard role-taking questions focused on the assessment of each role-taking level. For example:

Level 1—Subjective Role-Taking

(a) Does Holly know how Sean feels about the kitten? Why?

(b) Does Sean know why Holly cannot decide whether or not to climb the tree? Why or why not?

(c) Why might Sean think Holly will not climb the tree if Holly does not tell him about her promise?

Level 2—Self-Reflective Role-Taking
    (a) What does Holly think her father will think of her if he finds out?
    (b) Does Holly think her father will understand why she climbed the tree? Why is that?

Level 3—Mutual Role-Taking
    (a) What does Holly think most people would do in this situation?
    (b) If Holly and her father discussed this situation, what might they decide together? Why is that?
    (c) Do you know what the Golden Rule is [explain if child says no]? What would the Golden Rule say to do in this situation? Why?

    In addition, open-ended discussion in the Piagetian tradition (the Piagetian clinical method) and role-playing techniques were used further to assess level of role-taking.
    Green's index of reproducibility indicated that questions comprehended at each level formed a Guttman scale on both tasks ($I = .88$ and $.84$, respectively). Subjects were scored at the highest level of role-taking clearly exhibited. For example, if a S comprehended a level 3 question or used reasoning indicative of the level 3 structure in the open-ended part of the interview, he was scored at level 3.

RESULTS AND DISCUSSION

The correlation between highest level of role-taking attained in each of the measures was .93. Therefore, to define a subject's highest level, a clinical assessment was made over both situations to decide if the concept was clearly evident. Percentage perfect agreement on highest level attained between trained scorers was .96; percentage one level apart was .04. Differences were resolved upon discussion. The percentage perfect agreement across 20 randomly selected protocols between a trained and an untrained scorer, who used the role-taking scoring manual, without theoretical or experiential background, was .78.

Analysis indicated a significant product-moment coefficient of correlation of role-taking level of age, r (40) = .80; $p < .001$. No significant sex differences were found. Table 8.1 presents the percentage of Ss at each role-taking level by age.

The results of this study support our contention that social role-taking can be conceptually defined in structural terms. The sequence of structures, which was constructed on the basis of past theory and research, was found to emerge empirically in an age-related fashion. The age norms closely parallel those reported by Feffer for his system of levels. At our level 0, which is predominant among the 4-year-olds, there is no evidence of differentiation and therefore no coordination of perspectives. At level 1, which was present in most 6-year-olds' thinking, although a distinction is made between perspectives, Ss failed to coordinate them. This level is parallel to Feffer's level of simple focusing, in evidence at age 6, in which S manages to change perspective but without maintaining consistency. Our level 2 was the predominant emergent structure of the 8-year-olds. Perspectives at this level are taken in a sequential manner paralleling Feffer's level of consistent elaboration, also reaching a peak at ages 7 and 8. At our level 3, which was in evidence in the 8- and 10-year-old groups, perspectives are coordinated simultaneously, as in Feffer's level of change of perspective, beginning to emerge at age 9. The implication of this research is that role-taking structures can be identified within the context of moral dilemmas as well as in other interpersonal contexts and that the structures are similar in form and sequence to those described in other areas of interpersonal functioning.

There are several possible lines of future research that might clarify the nature of the sequence of structures we have defined. A first line of research is the examination of the structures in longitudinal studies over different populations to assess the degree to which they conform to the requirements of a true developmental sequence. Sec-

Table 8.1. Percentage of Subjects Reaching a Given Role-Taking Level at Each Level of Chronological Age ($N = 10$ per Age Group)

Stage	Age 4	Age 6	Age 8	Age 10
0	80	10	0	0
1	20	90	40	20
2	0	0	50	60
3	0	0	10	20
Total	100	100	100	100

ond, they may be examined in relation to development in the impersonal sphere. The egocentrism of stage 0 may have its counterpart in preoperational thought, the decentering of levels 1 and 2 may correspond to concrete operational ability, and the mutuality and infinite-regress character of level 3 might parallel the emergence of formal operations in the impersonal sphere.

REFERENCES

Baldwin, J. M. 1906. *Social and Ethical Interpretations in Mental Development.* New York: Macmillan.

Feffer, M. H. 1959. The cognitive implication of role-taking behavior. *Journal of Personality* 27:152–168.

—— 1971. Developmental analysis of interpersonal behavior. *Psychological Review* 77:197–214.

Feffer, M. H. and V. Gourevitch. 1960. Cognitive aspects of role-taking in children. *Journal of Personality* 28:383–396.

Flavell, J. H. 1968. *The Development of Role-Taking and Communication Skills in Children.* New York: Wiley.

Green, B. F. 1956. Attitude measurement. In G. Lindzey, ed., *Handbook of Social Psychology,* Cambridge, Mass.: Addison-Wesley.

Kohlberg, L. 1969. Stage and sequence: the cognitive-developmental approach to socialization. In D. Goslin, ed., *Handbook of Socialization Theory and Research.* Chicago: Rand-McNally.

Mead, G. H. 1934. *Mind, Self, and Society.* Chicago: University of Chicago Press.

Piaget, J. 1950. *The Psychology of Intelligence.* London: Routledge & Kegan Paul.

—— 1967. *Six Psychological Studies.* New York: Random House.

Selman, R. 1971. Taking another's perspective: role-taking development in early childhood. *Child Development* 42:1721–1734.

Selman, R. and D. Byrne. 1973. Manual for scoring stage of social role taking in moral and non-moral social interviews. Manuscript. Harvard University.

Childhood
Identity and the Self

Nine

The Process of Learning Parental and Sex-Role Identification

David B. Lynn

An extremely important component of the child's growing sense of identity is his or her identifications with parents and with other figures. An *identification* is the internalization of personality characteristics and role behaviors of another person (for example, one's father) or of a group of people (for example, men in general). In this selection, David Lynn distinguishes between the child's identification with parents and the more general identification with a sex role. Lynn suggests that the processes of learning parental and sex-role identification are different, and further, that these processes are different for males and females.

THE PURPOSE of this paper is to summarize the writer's theoretical formulation concerning identification, much of which has been published piece-meal in various journals. Research relevant to new hypotheses is cited, and references are given to previous publications of this writer in which the reader can find evidence concerning the earlier hypotheses. Some of the previously published hypotheses are considerably revised in this paper and, it is hoped, placed in a more comprehensive and coherent framework.

THEORETICAL FORMULATION

Before developing specific hypotheses, one must briefly define identification as it is used here. *Parental identification* refers to the inter-

107

nalization of personality characteristics of one's own parent and to unconscious reactions similar to that parent. This is to be contrasted with *sex-role identification*, which refers to the internalization of the role typical of a given sex in a particular culture and to the unconscious reactions characteristic of that role. Thus, theoretically, an individual might be thoroughly identified with the role typical of his own sex generally and yet poorly identified with his same-sex parent specifically. This differentiation also allows for the converse circumstances wherein a person is well identified with his same-sex parent specifically and yet poorly identified with the typical same-sex role generally. In such an instance the parent with whom the individual is well identified is himself poorly identified with the typical sex role. An example might be a girl who is closely identified with her mother, who herself is more strongly identified with the masculine than with the feminine role. Therefore, such a girl, through her identification with her mother, is poorly identified with the feminine role (Lynn 1962).

Formulation of Hypotheses

It is postulated that the initial parental identification of both male and female infants is with the mother. Boys, but not girls, must shift from this initial mother identification and establish masculine-role identification. Typically in this culture the girl has the same-sex parental model for identification (the mother) with her more hours per day than the boy has his same-sex model (the father) with him. Moreover, even when home, the father does not usually participate in as many intimate activities with the child as does the mother, e.g., preparation for bed, toileting. The time spent with the child and the intimacy and intensity of the contact are thought to be pertinent to the process of learning parental identification (Goodfield 1965). The boy is seldom if ever with the father as he engages in his daily vocational activities, although both boy and girl are often with the mother as she goes through her household activities. Consequently, the father, as a model for the boy, is analogous to a map showing the major outline but lacking most details, whereas the mother, as a model for the girl, might be thought of as a detailed map.

However, despite the shortage of male models, a somewhat stereotyped and conventional masculine role is nonetheless spelled out for the boy, often by his mother and women teachers in the absence of his father and male models. Through the reinforcement of the culture's highly developed system of rewards for typical masculine-role behavior and punishment for signs of femininity, the boy's early learned identification with the mother weakens. Upon this weakened mother identification is welded the later learned identification with a culturally defined, stereotyped masculine role.

(1) *Consequently, males tend to identify with a culturally defined masculine role, whereas females tend to identify with their mothers* (Lynn 1959).

Although one must recognize the contribution of the father in the identification of males and the general cultural influences in the identification of females, it nevertheless seems meaningful, for simplicity in developing this formulation, to refer frequently to *masculine-role identification* in males as distinguished from *mother identification* in females.

Some evidence is accumulating suggesting that (2) *both males and females identify more closely with the mother than with the father.* Evidence is found in support of this hypothesis in a study by Lazowick (1955) in which the subjects were 30 college students. These subjects and their mothers and fathers were required to rate concepts, e.g., "myself," "father," "mother," etc. The degree of semantic similarity as rated by the subjects and their parents was determined. The degree of similarity between fathers and their own children was not significantly greater than that found between fathers and children randomly matched. However, children did share a greater semantic similarity with their own mothers than they did when matched at random with other maternal figures. Mothers and daughters did not share a significantly greater semantic similarity than did mothers and sons.

Evidence is also found in support of Hypothesis 2 in a study by Adams and Sarason (1963) using anxiety scales with male and female high school students and their mothers and fathers. They found that anxiety scores of both boys and girls were much more related to mothers' than to fathers' anxiety scores.

Support for this hypothesis comes from a study in which Aldous and Kell (1961) interviewed 50 middle-class college students and their mothers concerning child-rearing values. They found, contrary to their expectation, that a slightly higher proportion of boys than girls shared their mothers' child-rearing values.

Partial support for Hypothesis 2 is provided in a study by Gray and Klaus (1956) using the Allport-Vernon-Lindzey Study of Values completed by 34 female and 28 male college students and by their parents. They found that the men were not significantly closer to their fathers than to their mothers and also that the men were not significantly closer to their fathers than were the women. However, the women were closer to their mothers than were the men and closer to their mothers than to their fathers.

Note that, in reporting research relevant to Hypothesis 2, only studies of *tested similarity, not perceived similarity,* were reviewed. To test this hypothesis, one must measure tested similarity, i.e., measure both the child and the parent on the same variable and compare the similarity between these two measures. This paper is not concerned with perceived similarity, i.e., testing the child on a given variable and then comparing that finding with a measure taken as to how the child thinks his parent would respond. It is this writer's opinion that much confusion has arisen by considering perceived similarity as a measure of parental identification. It seems obvious that, especially for the male, perceived similarity between father and son would usually be closer than tested similarity, in that it is socially desirable for a man to be similar to his father, especially as contrasted to his similarity to his mother. Indeed, Gray and Klaus (1956) found the males' perceived similarity with the father to be closer than tested similarity.

It is hypothesized that the closer identification of males with the mother than with the father will be revealed more clearly on some measures than on others. (3) *The closer identification of males with their mothers than with their fathers will be revealed most frequently in personality variables which are not clearly sex-typed.* In other words, males are more likely to be more similar to their mothers than to their fathers in variables in which masculine and feminine role behavior is not especially relevant in the culture.

There has been too little research on tested similarity between males and their parents to presume an adequate test of Hypothesis 3. In order to test it, one would first have to judge personality variables as to how typically masculine or feminine they seem. One could than test to determine whether a higher proportion of males are more similar to their mothers than to their fathers on those variables which are not clearly sex-typed, rather than on those which are judged clearly to be either masculine or feminine. To this writer's knowledge, this has not been done.

It is postulated that the task of achieving these separate kinds of identification (masculine role for males and mother identification for females) requires separate methods of learning for each sex. These separate methods of learning to identify seem to be problem-solving for boys and lesson-learning for girls. Woodworth and Schlosberg differentiate between the task of solving problems and that of learning lessons in the following way:

> With a problem to master the learner must explore the situation and find the goal before his task is fully presented. In the case of a lesson, the problem-solving phase is omitted or at least minimized, as we see when the human subject is instructed to memorize this poem or that list of nonsense syllables, to examine these pictures with a view to recognizing them later. (1954:529)

Since the girl is not required to shift from the mother in learning her identification, she is expected mainly to learn the mother-identification lesson as it is presented to her, partly through imitation and through the mother's selective reinforcement of mother-similar behavior. She need not abstract principles defining the feminine role to the extent that the boy must in defining the masculine role. Any bit of behavior on the mother's part may be modeled by the girl in learning the mother-identification lesson.

However, finding the appropriate identification goal does constitute a major problem for the boy in solving the masculine-role identification problem. When the boy discovers that he does not belong in the same-sex category as the mother, he must then find the proper sex-role identification goal. Masculine-role behavior is defined for him through admonishments, often negatively given, e.g., the mother's and teachers' telling him that he should not be a sissy without precisely indicating what he *should* be. Moreover, these negative admonishments are made in the early grades in the absence of male teachers to serve as models and with the father himself often unavailable as a model. The boy must restructure these admonishments in order to abstract principles defining the masculine role. It is this process of defining the masculine-role goal which is involved in solving the masculine-role identification problem.

One of the basic steps in this formulation can now be taken. (4) *In learning the sex-typical identification, each sex is thereby acquiring separate methods of learning which are subsequently applied to learning tasks generally* (Lynn 1962).

The little girl acquires a learning method which primarily involves (a) a personal relationship and (b) imitation rather than restructuring the field and abstracting principles. On the other hand, the little boy acquires a different learning method which primarily involves (a) defining the goal, (b) restructuring the field, and (c) abstracting principles. There are a number of findings which are consistent with Hypothesis 4, such as the frequently reported greater problem-solving skill of males and the greater field dependence of females (Lynn 1962).

The shift of the little boy from mother identification to masculine-role identification is assumed to be frequently a crisis. It has been observed that demands for typical sex-role behavior come at an earlier age for boys than for girls. These demands are made at an age when boys are least able to understand them. As was pointed out above, demands for masculine sex-role behavior are often made by women in the absence of readily available male models to demonstrate typical sex-role behavior. Such demands are often presented in the form of punishing, *negative* admonishments, i.e., telling the boy what not to do rather than what to do and backing up the demands with punishment. These are thought to be very different conditions from those in which the girl learns her mother-identification lesson. Such methods of demanding typical sex-role behavior of boys are very poor methods for inducing learning.

(5) *Therefore, males tend to have greater difficulty in achieving same-sex identification than females* (Lynn 1964).

(6) *Furthermore, more males than females fail more or less completely in achieving same-sex identification, but rather, they make an opposite-sex identification* (Lynn 1961).

Negative admonishments given at an age when the child is least able to understand them and supported by punishment are thought to produce anxiety concerning sex-role behavior. In Hartley's words:

> This situation gives us practically a perfect combination for inducing anxiety—the demand that the child do something which is not clearly defined to him, based on reasons he cannot possibly appreciate, and enforced with threats, punishments and anger by those who are close to him. (1959:458)

(7) *Consequently, males are more anxious regarding sex-role identification than females* Lynn 1964). It is postulated that punishment often leads to dislike of the activity that led to punishment (Hilgard 1962). Since it is "girl-like activities that provoked the punish-

ment administered in an effort to induce sex-typical behavior in boys, then, in developing dislike for the activity which led to such punishment, boys should develop hostility toward "girl-like" activities. Also, boys should be expected to generalize and consequently develop hostility toward all females as representatives of this disliked role. There is not thought to be as much pressure on girls as on boys to avoid opposite-sex activities. It is assumed that girls are punished neither so early nor so severely for adopting masculine sex-role behavior.

(8) *Therefore, males tend to hold stronger feelings of hostility toward females than females toward males* (Lynn 1964). The young boy's same-sex identification is at first not very firm because of the shift from mother to masculine identification. On the other hand, the young girl, because she need make no shift in identification, remains relatively firm in her mother identification. However, the culture, which is male-dominant in oreintation, reinforces the boy's developing masculine-role identification much more thoroughly than it does the girl's developing feminine identification. He is rewarded simply for having been born masculine through countless privileges accorded males but not females. As Brown pointed out:

> The superior position and privileged status of the male permeates nearly every aspect, minor and major, of our social life. The gadgets and prizes in boxes of breakfast cereal, for example, commonly have a strong masculine rather than feminine appeal. And the most basic social institutions perpetuate this pattern of masculine aggrandizement. Thus, the Judeo-Christian faiths involve worshipping God, a "Father," rather than a "Mother," and Christ, a "Son," rather than a "Daughter." (Brown 1958)

(9) *Consequently, with increasing age, males become relatively more firmly identified with the masculine role* (Lynn 1959).

Since psychological disturbances should, theoretically, be associated with inadequate same-sex identification and since males are postulated to be gaining in masculine identification, the following is predicted: (10) *With increasing age males develop psychological disturbances at a more slowly accelerating rate than females* (Lynn 1961).

It is postulated that as girls grow older, they become increasingly disenchanted with the feminine role because of the prejudices against their sex and the privileges and prestige offered the male rather than the female. Even the women with whom they come in contact are likely to share the prejudices prevailing in this culture against their own sex (Kitay 1946). Smith (1939) found that with increasing age girls have a progressively better opinion of boys and a progressively poorer opinion of themselves. (11) *Consequently, a larger proportion of females than males show preference for the role of the opposite sex* (Lynn 1959).

Note that in Hypothesis 11 the term "preference" rather than "identification" was used. It is *not* hypothesized that a larger proportion of females than males *identify* with the opposite sex (Hypothesis 6 predicted the reverse) but rather that they will show *preference* for the role of the opposite sex. *Sex-role preference* refers to the desire to adopt the behavior associated with one sex or the other or the perception of such behavior as preferable or more desirable. *Sex-role preference* should be contrasted with *sex-role identification*, which, as stated previously, refers to the actual incorporation of the role of a given sex and to the unconscious reactions characteristic of that role.

Punishment may suppress behavior without causing its unlearning (Hilgard 1962). Because of the postulated punishment administered to males for adopting opposite-sex role behavior, it is predicted that males will repress atypical sex-role behavior rather than unlearn it. One might predict then a discrepancy between the underlying sex-role identification and the overt sex-role behavior of males. For females, on the other hand, no comparable punishment for adopting many aspects of the opposite-sex role is postulated. (12) *Consequently, where a discrepancy exists between sex-role preference and identification, it will tend to be as follows: Males will tend to show same-sex role preference with underlying opposite-sex identification. Females will tend to show opposite-sex role preference with underlying same-sex identification* (Lynn 1964). Stated in another way, where a discrepancy occurs both males and females will tend to show masculine-role preference with underlying feminine identification.

Not only is the masculine role accorded more prestige than the feminine role, but males are more likely than females to be ridiculed or punished for adopting aspects of the opposite-sex role. For a girl to be a tomboy does not involve the censure that results when a boy is a sissy. Girls may wear masculine clothing (shirts and trousers), but boys may not wear feminine clothing (skirts and dresses). Girls may play with toys typically associated with boys (cars, trucks, erector sets, and guns), but boys are discouraged from playing with feminine toys (dolls and tea sets). (13) *Therefore, a higher proportion of females than males adopt aspects of the role of the opposite sex* (Lynn 1959).

Note that Hypothesis 13 refers to *sex-role adoption* rather than *sex-role identification* or *preference*. *Sex-role adoption* refers to the overt behavior characteristic of a given sex. An example contrasting sex-role adoption with preference and identification is an individual who *adopts* behavior characteristic of his own sex because it is expedient, not because he *prefers* it or because he is so *identified*.

SUMMARY

The purpose of this paper has been to summarize the writer's theoretical formulation and to place it in a more comprehensive and coherent framework. The following hypotheses were presented and discussed:

1. Males tend to identify with a culturally defined masculine role, whereas females tend to identify with their mothers.
2. Both males and females identify more closely with the mother than with the father.
3. The closer identification of males with their mothers than with their fathers will be revealed most frequently in personality variables which are not clearly sex-typed.
4. In learning the sex-typical identification, each sex is thereby acquiring separate methods of learning which are subsequently applied to learning tasks generally.
5. Males tend to have greater difficulty in achieving same-sex identification than females.
6. More males than females fail more or less completely in achieving same-sex identification but rather make an opposite-sex identification.
7. Males are more anxious regarding sex-role identification than females.
8. Males tend to hold stronger feelings of hostility toward females than females toward males.

9. With increasing age, males become relatively more firmly identified with the masculine role.
10. With increasing age, males develop psychological disturbances at a more slowly accelerating rate than females.
11. A larger proportion of females than males show preference for the role of the opposite sex.
12. Where a discrepancy exists between sex-role preference and identification, it will tend to be as follows: Males will tend to show same-sex role preference with underlying opposite-sex identification. Females will tend to show opposite-sex role preference with underlying same-sex identification.
13. A higher proportion of females than males adopt aspects of the role of the opposite sex.

REFERENCES

Adams, E. and I. Sarason. 1963. Relation between anxiety in children and their parents. *Child Development* 34:237–246.

Adlous, J. and L. Kell. 1961. A partial test of some theories of identification. *Marriage and Family Living* 23:15–19.

Brown, D. 1958. Sex-role development in a changing culture. *Psychological Bulletin* 55:235.

Goodfield, B. 1965. A preliminary paper on the development of the time intensity compensation hypothesis in masculine identification. Paper read at the San Francisco State Psychological Convention, April.

Gray, S. and R. Klaus. 1956. The assessment of parental identification. *Genetic Psychology Monographs* 54:87–114.

Hartley, R. 1959. Sex-role pressures and the socialization of the male child. *Psychological Reports* 5:458.

Hilgard, E. 1962. *Introduction to Psychology.* New York: Harcourt, Brace, and World.

Kitay, P. 1940. A comparison of the sexes in their attitudes and beliefs about women: a study of prestige groups. *Sociometry* 3:399–407.

Lazowick, L. 1955. On the nature of identification. *Journal of Abnormal and Social Psychology* 51:175–183.

Lynn, D. 1959. A note on sex differences in the development of masculine and feminine identification. *Psychological Review* 66:126–135.

—— 1961. Sex differences in identification development. *Sociometry* 24:372–383.

—— 1962. Sex-role and parental identification. *Child Development* 33:555–564.
—— 1964. Divergent feedback and sex-role identification in boys and men. *Merrill-Palmer Quarterly* 10:17–23.
Smith, S. 1939. Age and sex differences in children's opinion concerning sex differences. *Journal of Genetic Psychology* 54:17–25.
Woodworth, R. and H. Schlosberg. 1954. *Experimental Psychology*, p. 529. New York: Henry Holt.

Ten

Development of a Sense of Self-Identity in Children

Carol J. Guardo and Janis Beebe Bohan

While the major landmark in the identity and self domain during infancy is the development of the ability to recognize oneself (see selection 3), the major landmarks during childhood are the development of identifications (see selection 9) and of an increasingly sophisticated and differentiated *self-concept*, or *self-identity*. In this selection, Guardo and Bohan chart some important changes between the ages of six and nine in how children conceive of themselves. As do Selman and Byrne in their discussion of role-taking (selection 8), these authors relate changes in self-identity to the child's general cognitive development.

DEVELOPMENT OF SELF has been widely treated in the theoretical literature of psychology (Maslow 1962; May 1967; Rogers 1961) but subjected to negligible empirical investigation. The present research was designed to remedy this situation in part and had as its specific objectives: distinguishing between self and self-concept, redefining the theoretical construct of self and rendering it amenable to operational definition, and investigating the development of self or a sense of self-identity in children.

Often the distinction between self and self-concept has been made in theory, but in practice the constructs have been treated as synonymous. Wylie (1961) pointed out two chief meanings of self: self as subject (the individual as experiencer) and self as object (the individual as known to himself). Her review of the literature was concerned with the second meaning, that is, with self-concept, which typically has referred to the composite of ideas and perceptions that

118

the individual has about his abilities, accomplishments, faults, weaknesses, and values. Self-concept has been treated as a hypothetical construct, inferred from specified behaviors, and rendered quantitatively measurable.

In contrast, self (self as subject) may be conceived of as *that experience of which the self-concept is a concept.* It refers to an immediate, intra-organismic experience whereby the individual is aware of his own being and functioning. This experience is equated with the individual's sense of his own self-identity or with self as an experiential reality. Just as any other psychological concept can be treated as a hypothetical contrast, so too can self; it can be inferred from behavior (e.g., verbal report) and operationally defined.

Empirical research dealing with self and its development has been minimal. Several studies have taken the increasing use of personal pronouns with age as an indicant of self-emergence. Studies involving observation of one child were conducted by Moore (1896), Cooley (1908), and Bain (1936); those involving observations of large samples of children were conducted by Goodenough (1938) and Ames (1952). Wylie's criticism of Ames's work is germane to an evaluation of the research cited: "Procedures for standardizing recording conditions, exploring interobserver reliability, and demonstrating construct validity of behavior categories as indices of 'sense of self' were not followed" (Wylie 1961:119). The present research was designed to correct for some of these weaknesses, as well as to explore a new avenue for the investigation of self.

In order to render self operationally definable, the construct was more explicitly defined *"from the point of view of the experiencer* as a phenomenological feeling or sense of self-identity" (Guardo 1968:139). A sense of self-identity involves the experience of having or being certain characteristics which are essential to the human individual in that they contribute to "personeity," that is, to this experience of himself as a person with a unique identity. It was postulated that some critical dimensions of a sense of self-identity are the senses of humanity, sexuality, individuality, and continuity. These four dimensions were thought to be necessary, but not necessarily sufficient, characteristics of the sense of self-identity.

A sense of humanity refers to the individual's awareness that he possesses distinctively human potentialities and experiences. Sexuality consists of the individual's sense of his own maleness (or her own femaleness) whereby he identifies himself in terms of his sex and its behavioral implications. The third dimension, a sense of individuality, refers to the awareness on the part of the individual that he is a singular and unique being with an identity of his own. The last postulated dimension, continuity, is comparable to other theoretical notions (e.g., Calkins 1908; Erikson 1968) and refers to the individual's experience of his present self (the personal identity that he now is) as continuous with what he has been in the past and will become in the future. In summary, a sense of self-identity or "personeity" encompasses that experience whereby the individual is aware that he is one being with a unique identity who has been, is, and will be a male (or female) human person separate from and entirely like no other.

The objectives of the present research were to define operationally the postulated dimensions of a sense of self-identity and to ascertain whether these dimensions evidenced a developmental pattern of emergence. The specific expectations were that: the sense of self-identity would show a gradual unfolding across time, the postulated dimensions of a sense of self-identity would be demonstrable aspects of self-experience, the development evidenced in the child's sense of self-identity would be a basis for augmenting existing theories of self, and these developments would parallel the course of cognitive development charted in Piaget's theory (1966).

The last expectation was based on a methodological and a theoretical consideration. Procedurally, after each critical inquiry in the self-identity questionnaire, the child was asked to give a reason for his answer. Since, when asked for reasons, the child must exercise whatever cognitive functions he has, his explanations were expected to show the Piagetian course of cognitive development. On the theoretical side, an identity, according to Piaget (1968), involves the perception of qualitative (nonquantitative) invariants underlying a constancy. A sense of self-identity involves the perception of qualitative invariants (the postulated dimensions of self-identity) which underlie a constancy, namely, the fully crystallized sense of personal identity or personeity. Since Piaget contended that identity changes functionally from one cognitive stage to another in the manner in which the invariants are perceived or cognized, it was expected that the way children perceive self-identity would show changes paralleling the changes in the cognitive function of identity.

METHOD

Subjects

Subjects were 116 middle-class Caucasian pupils in kindergarten through third grade at a public school in a suburb of Rochester, New York. Table 10.1 shows a distribution of Ss by sex and age.

Measure

Extensive interview construction and piloting were done to develop a suitable methodology. A Piaget-type technique (Piaget 1965) was used consisting of a semistructured interview comprised of questions designed to probe the sense of self-identity (SSI). In the interview, S was not asked to report directly on his experience of self; instead, he was asked if he could assume an identity different from his own and yet not give up his own personal identity. Presumably, he would not recognize and could not handle the dilemma posed unless he had a sense of self-identity.

The questions designed to tap the four dimensions of SSI are described below.

1. *Humanity.* After E ascertained what kind of pet the child had or would like to have, S was asked if he could assume the identity of this pet (a nonhuman being). It was reasoned that if S had a sense of humanity, he would prescind from the possible sexual and behavioral similarities and cite the more essential impediment of his humanness.

2. *Sexuality.* After S had identified a sibling or peer of the opposite sex, he was asked if he could take on the identity of that

Table 10.1. Subject Numbers by Age and Sex

Age	Males	Females	Total
6	17	11	28
7	12	20	32
8	15	13	28
9	15	13	28
Total	59	57	116

other person. Here, humanity is shared and the more obvious impediment to the proposed identity is presumably the sexual difference.

3. *Individuality.* Subject identified a sibling or peer of the same sex and then was asked if he could assume the identity of this other person. In this case, both humanity and sexuality are shared and thus the critical hindrance to taking on another identity is the fact (or sense) of his own individuality.

4. *Continuity.* Three questions were asked about the sense of continuity—one each regarding continuity from the past, into the near future, and into the remote future. These were designed to assess whether S sensed that in some way he had always been the same person since his birth, and that he would always be the same person in both the near future ("the higher grades") and the remote future (when "grown up"). . . .

Interview Procedure

Each S was escorted individually into a private office in the school. The taped interviews conducted by one of us (J. B. B.) lasted 15–20 minutes. The tapes served as the basis for subsequent coding and categorizing of responses. Each S was assured that there were no right or wrong answers to the questions and was encouraged to explain his answers fully. The same order of questioning was followed by all Ss.

There was no age ordering in the sequence of interviews; for example, some kindergartners were interviewed early in the sessions and some later. This procedure was followed in order to balance out any effect which might have resulted from a change in E's approach across time.

Scoring Procedures

Three response categories were used for scoring answers to all questions except that dealing with continuity from the past. The categories were yes, no, or conditional (which included responses where Ss cited some qualification of the if-then variety for their answers). For the question concerning continuity into the past, four categories of response were used (see Results section).

The reasons which Ss gave for their answers to each of the critical questions were placed into 11 categories. These were derived from a study of the rationales given by the pilot *S*s and from a random sampling of the taped responses from the study proper. The categories included, for example, differences or similarities in physical characteristics, age, sex, mental functions, and overt behaviors.

Each of the authors independently coded all responses from the tapes and categorized them into the aforementioned categories. The agreement in judgments for all age-sex groups of *S*s ranged from 93.1 to 100 percent.

RESULTS AND DISCUSSION

Chi-square values and contingency coefficients (see table 10.2) based on age of *S*s and category of response were calculated for each of the SSI dimensions for both sexes. Thirteen of the χ^2s were significant. Only one (for males regarding continuity into the near future) was nonsignificant. However, on the basis of these findings alone, it could not be concluded that a sense of self-identity is a developmental phenomenon, even though these findings showed that age and SSI were related. This relationship is complex and can only be interpreted by drawing on two different sets of data—the quantitative (the percentages of *S*s answering in given ways) and the qualitative (the kinds of reasons *S*s gave for their answers). If only the quantitative data (in table 10.2) were considered, then it would appear, in some cases, that younger *S*s (ages 6 and 7) were more in accord with the theoretically anticipated results than older *S*s (ages 8 and 9). Moreover, looking across SSI dimensions, it could be inferred that most of the children of the ages studied had developed some sense of self-identity, with no implication that the SSI of older *S*s was any more developmentally advanced than that of younger *S*s. However, a closer examination of the quantitative data indicated that older *S*s gave more conditional responses than did younger *S*s—responses which are developmentally more sophisticated than the simple affirmative or negative responses of the younger *S*s. And further, an analysis of the qualitative data or the rationales provided for responses showed that the SSI of the older *S*s was qualitatively different from and more fully developed than that of younger *S*s. On these combined bases, it was concluded that SSI is a developmental phenomenon and much of the

Table 10.2. χ^2s, C Coefficients, and Percentages of Response by Age, Sex, and Response Category for Each Self-Identity Dimension

	Humanity			Sexuality			Individuality			Continuity-P				Continuity-NF			Continuity-RF			Name		
Age	Yes	No	Con	Yes	No	Con	Yes	No	Con	DK	TS	TL	AL	Yes	No	Con	Yes	No	Con	Yes	No	Con
Girls																						
6	0.0	100.0	0.0	0.0	100.0	0.0	9.1	81.8	0.0	72.7	18.2	0.0	9.1	81.8	9.1	0.0	63.6	27.3	0.0	63.6	36.4	0.0
7	10.0	85.0	5.0	5.0	85.0	10.0	15.0	75.0	10.0	35.0	15.0	5.0	40.0	75.0	20.0	5.0	60.0	30.0	10.0	35.0	25.0	40.0
8	7.7	76.9	15.4	7.7	69.2	23.1	23.1	30.8	46.1	15.4	23.1	0.0	61.5	84.6	7.7	7.7	76.9	7.0	15.4	84.6	7.7	7.7
9	15.4	76.9	7.7	23.1	46.1	30.8	7.7	53.8	38.5	15.4	15.4	0.0	69.2	84.6	7.7	7.7	61.5	23.1	15.4	61.5	0.0	30.7
	$\chi^2 = 36.66$			$\chi^2 = 89.62$			$\chi^2 = 99.50$			$\chi^2 = 127.87$				$\chi^2 = 17.52^*$			$\chi^2 = 31.51$			$\chi^2 = 120.91$		
	$C = .49$			$C = .66$			$C = .68$			$C = .72$				$C = .36$			$C = .46$			$C = .71$		
Boys																						
6	17.7	82.3	0.0	5.9	82.3	11.8	11.8	58.8	23.5	29.4	17.7	23.5	29.4	94.1	5.9	0.0	64.7	35.3	0.0	64.6	17.7	17.7
7	8.3	91.7	0.0	0.0	91.7	8.3	0.0	75.0	25.0	16.7	50.0	0.0	33.3	91.7	8.3	0.0	25.0	58.3	16.7	33.3	41.7	25.0
8	13.3	66.7	20.0	6.7	73.3	20.0	20.0	40.0	40.0	13.3	40.0	0.0	46.7	93.3	6.7	0.0	73.3	13.3	6.7	60.0	0.0	33.3
9	26.6	73.3	0.0	0.0	80.0	20.0	26.6	46.8	26.6	6.7	20.0	0.0	73.3	93.3	6.7	0.0	66.7	20.0	13.3	60.0	13.4	26.6
	$\chi^2 = 75.61$			$\chi^2 = 21.82^*$			$\chi^2 = 44.98$			$\chi^2 = 87.85$				$\chi^2 = 0.51$			$\chi^2 = 81.59$			$\chi^2 = 65.88$		
	$C = .63$			$C = .40$			$C = .53$			$C = .66$				$C = .07$			$C = .64$			$C = .60$		
														(N.S.)								

NOTE. Some totals for a given age do not equal 100% because some Ss gave no response. All χ^2 are significant at $p > .001$, except those shown with asterisk ($^*p > .01$). Con = conditional response; continuity-P = continuity from the past; continuity-NF = continuity into the near future; continuity-RF = continuity into the remote future; DK = "don't know"; TS = "too short"; TL = "too long"; AL = "always."
$^*p > .01$.

124

interpretation of the differences between the SSI dimensions of younger and older children pivoted on a consideration of the reasons *S*s gave for their answers.

Humanity
    The dimension of humanity was assessed by asking *S*s if they could take on the identity of an animal (nonhuman) being. Most *S*s gave negative responses (see table 10.2), as anticipated. When younger *S*s answered negatively, they generally cited perceptual and/ or behavioral dissimilarities as the impediment to the nonhuman identity. In contrast, older *S*s typically responded by matter-of-factly stating that they could not be animals because they were humans or possessed human characteristics (e.g., the ability to talk or think). Although some older *S*s only asserted their humanity, those who gave reasons had recourse to more subtle and generally covert differences as opposed to the externally observable differences noted by younger *S*s. The differences between the responses of younger and older *S*s can be seen in the contrast between these two examples.[1]

    F 5.5: He's brown and black and white. He has brown eyes. 'Cause he walks like a dog.
    F 7.6: He's an animal and I'm not. I'm a human being. A human being can talk and has arms and legs and they can learn more things.

Sexuality
    The pattern of responses concerning sexual identity is complex. Girls gave a decreasing percentage of negative (anticipated) answers with age and an increased percentage of conditional replies. Boys at all ages gave more negative responses than did girls, and older boys gave more conditional responses than did younger boys.
    Younger *S*s (ages 6 and 7) of both sexes generally cited perceptual appearance or behavioral dissimilarities as impediments to the assumption of the identity of the opposite sex. Most simply as-

---

[1] For this and other verbatim examples to follow, the sex of S is indicated as (M) or female (F) and followed by age in years and months.

serted that they were a member of one sex, could not be a member of the opposite sex, and did not explain why this fact alone constituted an essential impediment to the proposed identity. Older Ss also cited differences in appearance and behavior, but those who answered conditionally argued that, if these differences could be overcome (e.g., by an act of God), then they could assume the identity of an opposite-sexed person.

Older boys who answered negatively responded like younger boys by asserting that they were members of one sex and simply could not assume the opposite sexual identity. In addition, of the few who cited personeity as the basic impediment to the identity change, most were girls. For example:

F 8.10: Because I'm me and he's him. Well, that I can't change in any way 'cause I've got to stay this, like myself.

The notion of personeity requires greater abstracting ability than does the citing of behavioral and perceptual differences. Thus, these findings, coupled with the girls' increase in conditional responses, may indicate that the girls sampled were more advanced cognitively than the boys or that they were more sensitive to individuality than to sexuality.

Individuality

Responses to the question regarding individuality yielded age and sex differences (see table 10.2). Eight-year-olds of both sexes gave the smallest percentage of negative (anticipated) answers and the highest percentage of conditional replies. Younger and older Ss differed in the rationales provided for their answers, and girls of all ages gave more personeity reasons than did boys (although the total number of such rationales was small).

The minority of Ss who said that they could assume the identity of another person argued that perceptual and behavioral similarities provided the basis of the proposed identity switch. Older Ss did not argue, however, as had a few younger Ss, that their own volition was sufficient cause for the change. This finding corresponds with Piaget's contention (1954) that psychological causality (the sense of causing one's own actions through willing) decreases with age in children.

For negative answers, many older and younger Ss cited differences in appearance and behavior as impediments to the alternate identity. Some older Ss, however, cited covert differences in feelings, thoughts, and interests.

Younger boys who gave conditional responses used the mere elimination of behavioral and perceptual dissimilarities as a basis for a possible identity switch, but older Ss speculated about an external agency which could effect the identity change.

> M 8.10: It's not possible. . . . I don't know. . . . Well, if there was a machine that you could do it with, it would be possible.

At all ages and for both sexes, some few argued for their immutability as a person. They maintained that they had been made one person, would always be that same person, and thus could be no other.

> F 7.2: No, 'cause I'm already a person—you can't change 'cause one person is enough. You can only be one person 'cause you can't change.

Considering the small number of Ss who asserted personeity, it might be concluded that most children age 9 or younger generally have not grasped the postulated meaning of individuality. However, an analysis of the increased percentages of conditional responses of older children (see table 10.2) qualifies this conclusion. Older Ss confronted the possibility of an alternate identity while younger Ss generally did not conceive of alternatives. Seemingly, the younger (preoperational) child differentiates between himself and other beings on the basis of observable behaviors and characteristics. As Piaget (1966) has contended, the preoperational child deals with what is perceptually apparent, while the older (concrete-operational) child also conceives of *possibles*—in this case, possible identities. His recognition of multiple possibilities then raises the question of the probability of each, but the notion of probability remains incomplete during the concrete-operational period (Davies 1965). Hence, the child may argue that, given certain improbable conditions, he might possible assume another identity as in the example (M 8.10) above.

Continuity from the Past (Continuity-P)

In order to ascertain whether Ss had a sense of continuity from the past, they were asked how long they had been the boy (or girl) they now were. When S's reply indicated a period of time shorter than his chronological age, the response was labeled "too short" (TS). Boys gave more of these responses than girls, with the attendant rationale being that they were different when they were babies or before they went to school. Subjects who provided answers which were "too long" (TL) stated that they had been the same for periods of time longer than their chronological age. These responses came primarily from 6-year-old boys.

In general, as table 10.2 indicates, both sexes gave decreasing percentages of "don't know" (DK) responses and increasing numbers of "always" (AL) responses with age. However, since Ss gave little explanation for their replies to this question, it is only cautiously concluded that the increase with age in AL responses points up the emergence of a sense of continuity from the past.

Continuity into the Near Future (Continuity-NF)

Generally, a majority of Ss of both sexes at all ages answered affirmatively or as anticipated. Their reasons centered mainly around the retention of the same name and physical characteristics, despite anges brought about by physical growth. Older Ss (ages 8 and 9) asserted that they would remain the same not only despite physical growth, but also despite the fact that they would "know more," be "smarter"—that is, despite cognitive growth.

Again, there was a small subgroup of Ss who gave rationales involving personeity. There were no obvious differences in their frequency according to sex or age.

The fact that many Ss cited their names as a basis for Continuity-NF (as well as for Continuity-P) supports the notion of nominalism in children (Piaget 1965). Older Ss tended to give other reasons for continuity along with mention of their names; younger Ss relied more heavily on their unchanging names as the basis for sameness across time. This tendency on the part of younger children to resort to nominal realism in their sense of self-identity was confirmed by the results derived from the question concerning the removal of their names. Considerably larger percentages (see table 10.2) of younger as compared with older Ss argued that they would not be the same persons if their names were taken away from them. By the age of 8, however, the recognition that names are arbitrary designations and not guarantees of identity was clear-cut. As one precocious S put it:

F 6.5: Because my name isn't what I am. It's just my name!

Continuity into the Remote Future (Continuity-RF)

A sense of Continuity-RF was implied by Ss' contentions that they would be the same even when they became grown-ups. Although the majority of Ss answered as anticipated, boys age 7 gave proportionately more negative than affirmative responses. This finding is clarified by contrasting their replies with those of both younger Ss and girls age 7.

Six-year-olds of both sexes and girls age 7 who answered negatively tended to key on physical growth changes, but boys age 7 gave reasons for discontinuity based on behavioral changes and changes in *role*. They stated that they would not be the same when they became grown-ups because they would be men (vs. boys), married, and working. Girls, however, did not begin to mention role change as a basis for discontinuity until the age of 8. The girl of 7 typically tried to deal with the change in name entailed by marriage rather than with the role change.

There was again a small group of Ss of both sexes and at all ages who cited personeity as a basis for continuity. For example:

F 9.0: Cause I'll just be the same. I don't change anything. [E: What will be the same?] Me. I just don't turn into anything. I just stay the same and stuff, and I grow.

Overall, the majority of Ss of both sexes at all ages gave evidence of a sense of Continuity-RF. Continuity-RF appears to be less certain, however, than Continuity-NF, especially when the implications of the change of role from child to adult are considered. Another factor contributing to the decrease in certainty about Continuity-RF may be an increased nebulousness of the more distant future.

In summary, a capsule description of the SSI of each age group studied follows. Children age 6 recognized that they are distinct sexual, human individuals in terms of their physical appearance and behavioral capacities. Nominal realism often served as a basis for their continuity into the future, but continuity from the past was generally confusing. The SSI of 7-year-olds was similar to that of 6-year-olds, although continuity from the past was more frequently recognized.

Children age 8 and 9 presented a contrasting picture. The increase in conditional responses showed that these Ss were considering possible, concrete alternatives to the dimensions of their own identities. The sexual, human, and individual distinctiveness recognized was based not only on the obvious differences in physical appearance and behavior, but also on the more covert and personalized differences in feelings, attitudes, and mediated behaviors. The larger percentage of personeity responses made by the older Ss seemed to indicate that SSI was becoming more crystallized than that of younger Ss. Sex differences were more apparent in the responses of the older Ss and can be summarized by noting that for girls "individuality" seemed to be the key dimension of SSI, while the boys "sexuality" or their masculinity was the key dimension.

Of all the Ss, about one child in six (mostly girls) recognized that name, physical characteristics, behaviors, and so on, are not the essential anchor of identity but, rather, personeity, the feeling of being a singular and personal identity, is. Personeity impedes the assumption of another human individuality and provides for continuity across time. This realization can perhaps be summarized by an existential rephrasing of the Cartesian first principle—"I am; therefore, I am I." As one child, who had given personeity responses to all the questions, expressed herself when asked if she would remain the same if her name were taken away:

F 8.10: I'd still be the same person. [E: What stays the same?]
         My per-son-al-i-ty?

In the last analysis, results based on the frequency of response supported the basic notion that children have a sense of self-identity showing age and sex differences. Results based on the reasons given for responses indicated that there are qualitative differences in the sense of self-identity of younger and older children. Taken collectively, these results generally confirmed the stated expectations of the investigation: (1) The sense of self-identity is a developmental phenomenon. (2) The implications of the age and sex differences found in the sense of self-identity point to ways in which existing self-theory can be revised and refined, although space does not permit elaboration of these here. (3) The dimensions of a sense of self-identity were found to vary qualitatively as a function of age (or age-related variables). (4) These qualitative differences can be interpreted in terms derived from Piaget's theory of cognitive development; specifically, the sense of self-identity can logically and cogently be treated as an identity involving qualitative constancies.

REFERENCES

Ames, L. B. 1952. The sense of self of nursery school children as manifested by their verbal behavior. *Journal of Genetic Psychology* 81:193–232.

Bain, R. 1936. The self-and-other words of a child. *American Journal of Sociology* 41:767–775.

Calkins, M. W. 1908. Psychology as science of self, II: The nature of the self. *Journal of Philosophy, Psychology and Scientific Methods* 5:64–68.

Cooley, C. H. 1908. A study of the early use of self-words by a child. *Psychological Review* 15:339–357.

Erikson, E. H. 1968. *Identity: Youth and Crisis.* New York: Norton.

Goodenough, F. L. 1938. The use of pronouns by young children: a note on the development of self-awareness. *Journal of Genetic Psychology* 52:333–346.

Guardo, C. J. 1968. Self revisited: the sense of self-identity. *Journal of Humanistic Psychology* 8:137–142.

Maslow, A. H. 1962. *Toward a Psychology of Being.* New York: Van Nostrand.

May, R. 1967. *Psychology and the Human Dilemma.* New York: Van Nostrand.

Moore, K. C. 1896. The mental development of a child. *Psychological Monographs* 1:No. 3.

Piaget, J. 1954. *The Construction of Reality in the Child.* New York: Basic Books.

Piaget, J. 1965. *The Child's Conception of the World.* Paterson, N.J.: Littlefield, Adams.

—— 1966. *Psychology of Intelligence.* Paterson, N.J.: Littlefield, Adams.

—— 1968. *On the Development of Memory and Identity.* Barre, Mass.: Clark University Press.

Rogers, C. R. 1961. *On Becoming a Person.* Boston: Houghton Mifflin.

Wylie, R. C. 1961. *The Self Concept.* Lincoln: University of Nebraska Press.

Childhood

Relations with Others

Eleven

Peer Interaction and the Behavioral Development of the Individual Child
Willard W. Hartup

What is the contribution of interaction and relationships with peers to the child's development? In an earlier selection (#7), we examined the role of play with peers in the child's cognitive development. In this selection, the focus is more on the function of peer relations for social development and mental health. Willard Hartup argues that "access to age-mates, acceptance by them, and constructive interactions with them" seem to be "essential to normal social development." In particular, peers seem to play an especially important role in the development of social effectiveness, in the socialization of aggression, and in sexual socialization.

EXPERIENCE WITH PEERS is commonly assumed to make numerous contributions to child development. Such experiences are believed to provide a context for sex-role learning, the internalization of moral values, the socialization of aggression, and the development of cognitive skills. The research literature, however, contains relatively little hard data concerning the functional contributions of peer interaction to the development of the individual child. There is little evidence that the give-and-take occurring during peer interaction actually determines the moral structuring that occurs in middle childhood, as Piaget (1932) suggested; there is no direct evidence that rough-and-tumble play contributes to the effectiveness with which the human child copes with aggressive affect (Harlow 1969); and the contributions of peer attachments to social and intellectual development are largely unspecified.

135

Nevertheless, the purpose of this paper is to argue that peer interaction is an essential component of the individual child's development. Experience with peers is not a superficial luxury to be enjoyed by some children and not by others, but is a necessity in childhood socialization. And among the most sensitive indicators of difficulties in development are failure by the child to engage in the activities of the peer culture and failure to occupy a relatively comfortable place within it.

PEER RELATIONS AND THE INDIVIDUAL CHILD'S DEVELOPMENT

Attachment and Sociability

Recent research confirms that during the third and fourth years of life there is a decrease in the frequency with which the child seeks proximity with the mother (Maccoby and Feldman 1972), an increase in the frequency of attention-seeking and the seeking of approval relative to the frequency with which the child seeks affection (Heathers 1955), and a change in the objects toward whom social overtures are made; specifically, there is an increase in the frequency of contact with peers (Heathers 1955). Peer attachments become even more characteristic of the child's social life during middle childhood. The bonds that children establish with age-mates are dissimilar to the earlier bonds that are forged between mother and child (Maccoby and Masters 1970). First, children employ different behaviors to express affection to age-mates and to adults. They follow one another around, giving attention and help to each other, but rarely express verbal affection to age-mates, hug one another, or cling to each other. Moreover, children do not seem to be disturbed by the absence of a specific child even though the absence of specific adults may give rise to anxiety and distress.[1] Second, the conditions that elicit attachment activity differ according to whether the available attachment object is an adult or another child. For example, fear tends to elicit running to the teacher rather than fleeing to one's peers. Third, the

[1] A possible exception to this statement is the mild depression that is sometimes reported when younger children are separated from their siblings.

behavior of adults toward children differs qualitatively from the behavior of one child toward another. Adults do not engage in sustained periods of playful behavior with children; rather, they assume roles as onlookers or supervisors of children's playful activities. In fact, play appears to emerge in the human repertoire almost completely within the context of peer interaction.[2] This fact casts further doubt on the assumption that the parent-child and child-child social systems are manifestations of a unitary "attachment" orientation.

What developmental benefits does the child derive from peer attachments? What are the correlates and/or consequences of sociability with age-mates? What attitudes and orientations typify the child who is *not* involved in easygoing social activities with peers? To my mind, the best evidence (although not the only evidence) bearing on this problem is to be found in Bronson's (1966) analysis of the data from the Berkeley Guidance Study. Among three "central" behavioral orientations emerging in her analysis, the clearest was a bipolar dimension labeled *reserved-somber-shy/expressive-gay-socially easy*. Across each of four age periods, covering the ages from five through sixteen years, this orientation (social reservedness) in boys was associated with: (a) an inward-looking social orientation, (b) high anxiety, and (c) low activity. In later childhood, the correlates of reservedness also came to include: (a) vulnerability, (b) lack of dominance, (c) nonadventuresomeness, and (d) instability. In other words, lack of sociability in boys was correlated with discomfort, anxiety, and a general unwillingness to engage the environment.

The correlates of social reservedness among girls were not substantially different. The associations between sociability and vulnerability, level of activity, and caution were significant across all four age periods, and a correlation with passivity tended to increase over time. Note that the cluster of traits that surrounds low sociability may be indicative of behavior that is in greater accord with social stereotypes among the girls than among the boys. In general, though, the findings suggest that failure to be involved with one's peers is ac-

---

[2] Just why adults do not engage in long periods of play with their children is something of a mystery. From a psychological viewpoint, it is probably not possible for an adult to regress cognitively to a degree sufficient to permit sustained, childlike play. From an evolutionary viewpoint, it is probably not conducive to species survival for adult members of a troupe to spend large blocks of time playing with their offspring instead of hunting and gathering. The fact remains, however, that play occurs primarily in the context of peer interaction rather than in interaction with adults.

companied by a lower level of instrumental competence (Baumrind 1972) and higher anxiety than is the condition of high peer involvement. Other research provides a composite picture of the socially *rejected* child that is very much like the composite picture of the socially *inactive* child: He is neither outgoing nor friendly; he is either very high or very low in self-esteem; he is particularly dependent on adults for emotional support; he is anxious and inappropriately aggressive (Hartup 1970).

Although the foregoing findings are relatively firm, interpretation of them is difficult. Everything we know concerning the developmental consequences of peer involvement is based on correlational data. It is not clear, therefore, that peer involvement is instrumental in producing an outgoing, active, nonanxious, assertive posture toward the world, or whether a reverse interpretation is to be preferred. More than likely, some set of external influences is responsible for this whole configuration of traits, i.e., for both sociability and its correlates. But what might such external influences be? Clearly, biological factors may be operative as well as social influences. Individual differences in sociability stabilize early, such differences are not closely associated with child-rearing practices (Bronson 1966), and these characteristics possess moderate heritabilities (Scarr 1969). Thus, the origins of the linkage between general personal—social effectiveness and the presence or absence of effective peer relations remain obscure. It is important, however, for both theoreticians and practitioners to know that this linkage exists and to know that it characterizes child behavior across a variety of samples, times, and circumstances.

Aggression

Scattered evidence suggests that children master their aggressive impulses within the context of the peer culture rather than within the context of the family, the milieu of television, or the culture of the school. Nonhuman primate studies demonstrate rather convincingly that peer contact during late infancy and the juvenile stage produces two effects on the individual: he acquires a repertoire of effective aggressive behaviors; and he acquires mechanisms for coping with the affective outcomes of aggressive interaction (Harlow 1969). In fact, socialization seems to require both rough-and-tumble play and experiences in which rough play escalates into aggression, and deescalates into playful interaction (Hamburg and Van Lawick-Goodall 1974). Field studies suggest that such experiences are readily available to the young in all primate species, including homo sapiens, although opportunities are greater for males than for females (Jay 1968; Hartup 1974).

Whether parents can produce such a marked impact on the development of aggression is doubtful. The rough-and-tumble experiences necessary for aggressive socialization seem to be incompatible with the demands of maternal bonding because, for all primate species, some tie to the mother must be maintained after the time when socialization of aggression is begun. Fathers may contribute significantly to aggression learning, both because they provide frequent and effective displays of aggressive behavior with their children and because bonding to the father is "looser," more "secondary," and less constraining than is bonding with the mother. Selected studies support this: Research on father absence shows that boys from such homes are less aggressive than boys in father-present homes (Hetherington and Deur 1972). Nevertheless, whether fathers alone could effectively socialize their children's aggression—even their male children—remains doubtful. The father's social role in western culture requires him to spend most of his time outside the family and, even in close-knit family cultures, paternal contacts with the young child are insufficient to produce all of the learning required for the successful modulation of aggressive behavior.

Thus, nature seems to have prepared for human socialization in such a way that child—child relations are more important contributors to the successful control of aggressive motivation than parent-child relations. Patterson and his associates (Patterson, Littman, and Bricker 1967; Patterson and Cobb 1971) have confirmed convincingly the various ways in which reinforcement for both aggression and yielding to aggression are provided within the context of peer interactions.

According to this line of reasoning, children who show generalized hostility and unusual modes of aggressive behavior, or children who are unusually timid in the presence of aggressive attack, may be lacking exposure to certain kinds of contacts with peers, i.e., rough-and-tumble play. In other words, peer contacts that never allow for aggressive display or that allow only for successful aggression (never for unsuccessful aggression) may be precursors of malfunctioning in the aggression system. Clearly, this hypothesis is plausible when applied to boys, although it is more tenuous for girls. Traditional socialization produces women who are ineffectively prepared for exposure to aggressive instigating events (except, perhaps, for threats to their children). Women are notably more anxious and passive than men when exposed to aggressive instigation, and this sex difference may be greater than is good for the future of the species. In any event, if women are to assume social roles more like those of men, and men are to assume roles more like those of women, some manipulation of early peer experiences is necessary. Opportunities for early exposure to rough-and-tumble play must be as equal for males and for females as opportunities for exposure to other normative behaviors.

Sex

If parents were to be given sole responsibility for the socialization of sexuality, homo sapiens would not survive. With due respect to the efforts of modern sex educators, we must recognize that the parent-child relationship is no better suited to the task of socializing sex than to the task of socializing aggression. Evolution has established the incest taboo (Lindzey 1967), a taboo that is so pervasive that interaction with age-mates is virtually the only opportunity available to the child in which he may engage in the trial and error, the modeling, and the information-gathering that ultimately produce his sexual life-style.

There can be little doubt that sexual attitudes and the basic sexual repertoire are shaped primarily by contacts with other children. Kinsey (1948) said:

Children are the most frequent agents for the transmission of the sexual mores. Adults serve in that capacity only to a smaller extent. This will not surprise sociologists and anthropologists, for they are aware of the great amount of imitative adult activity which enters into the play of children, the world around. In this activity, play though it may be, children are severe, highly critical, and vindictive in their punishment of a child who does not do it "this way," or "that way." Even before there has been any attempt at overt sex play, the child may have acquired a considerable schooling on matters of sex. Much of this comes so early that the adult has no memory of where his attitudes were acquired. (p. 445)

These comments are repeated, nearly word for word, in the records of those many investigators who have observed the various nonhuman primate species in the field (Jay 1968).

Of course, the child's earliest identification of itself as male or female and the earliest manifestations of sex-typed behavior patterns derive from interactions with its parents. Parents are known to respond differentially to boys and to girls from infancy onward (Rothbart and Maccoby 1966), and the sex-typed outcomes of parent-child interaction have been discussed extensively in the child development literature (e.g., Lynn 1969). Nevertheless, there is strong evidence that the peer culture supports and extends the process of sex-typing beginning in the earliest preschool years. Sex is the overriding polarizer in peer group formation in all primate species from the point of earliest contact. Sex is a more powerful determinant of "who plays with whom" than age, race, social class, intelligence, or any other demographic factor with the possible exception of propinquity. And, clearly, this sex cleavage is instrumental in transmitting normative sex-role standards to the child. How else to account for the vast number of sex differences that have been observed in the social activities of children (e.g., aggression) beginning with the preschool years?. . .

Moral Development

According to Piaget (1932), both the quantity and the quality of social participation are related to the child's moral development. Moral understanding is assumed to derive partly from the amount of social interaction in which the child participates and partly from his centrality in the peer group. During early childhood, the child's behavior reflects an "objective" moral orientation (i.e., he believes that rules are immutable and the power of adults is absolute). Adoption of a "subjective" moral orientation requires some opportunity to view moral rules as changeable products of group consensus. For this purpose, social give-and-take is required. Such opportunity is not common in the child's experiences with his parents and teachers because social systems such as the family and the school are structured in authoritarian terms. Only in rare instances is there sufficient reciprocity in adult–child interaction to facilitate the disequilibration that is necessary to form a mature moral orientation. The peer group, on the other hand, is seldom organized along authoritarian dimensions and possesses the inherent characteristics for furthering moral development.

Precious little data exist to support the thesis that the *amount* of age-mate contact is associated with advanced moral development. Keasey (1971) has recently published a study, however, based on 144 preadolescents, in which he found that children who belonged to relatively more clubs and social organizations had higher moral judgment scores than children belonging to fewer organized groups. These results, which were somewhat stronger for boys than for girls, stand alone in revealing a correlation between amount of social participation with peers and advances in moral functioning.

Evidence is more profuse concerning the association between the *quality* of a child's peer contacts and the level of his moral reasoning. Keasey (1971) also reported that self-reports of leadership functions, peer ratings of leadership and friendship nominations, and teacher ratings of leadership and popularity were all positively related to level of moral judgment as assessed by Kohlberg's (1958) techniques. Irwin (1967) was unable to demonstrate consistent relations between popularity and measures of moral understanding within several groups of nursery school children, either because the samples were too homogeneous, or because the period in question is too early for the relation between acceptance and morality to be reflected in a systematic manner. Other data, though, support Keasey's findings: Gold (1962) reported that peer leaders have "more socially integrative" ideologies than nonleaders; Porteus and Johnson (1965) showed that acceptance was related to good moral judgment as perceived by peers; Campbell and Yarrow (1961) found that popular children, as compared with less popular children, made more extensive use of subtle inferences concerning the causes of other children's behavior; and Klaus (1959) found that accepted children tend to emphasize being neat and tidy, being a good sport, and being able to take a joke in their descriptions of classmates. Each of these diverse findings suggests that popularity is linked with effectively internalized social norms. Studies of peer group leaders also show them to be actively and appropriately sociable.

Once again, there is difficulty in interpreting the findings. Existence of a significant correlation between the extent of social participation and level of moral functioning does not prove that the latter derives from the former. Membership in clubs may facilitate change in the structuring of moral understanding, but higher levels of moral understanding may also be prerequisite to membership in large numbers of social clubs. Children may choose friends and teachers may nominate children as popular who can demonstrate advanced levels of moral understanding; on the other hand, popularity itself may enhance the child's level of moral reasoning. Another possibility is that the linkages cited here exist partly because estimates of moral understanding and estimates of popularity are both modestly related to intelligence.

Thus, the basic hypothesis that peer interaction contributes to advances in moral development remains unproven. The available evidence is consistent with this hypothesis, to be sure, but only controlled manipulation of the child's social experiences with age-mates can provide adequate causal evidence. Needless to say, such manipulations are very hard to produce.

Are children who have successful peer relations more overtly honest and upright than children who are loners and/or who are rejected by their peers? Some of the previously cited evidence is suggestive of such a state of affairs, although additional evidence is not extensive. In one early investigation, Roff (1961) found that in a sample of servicemen, all of whom were former patients in a child-guidance clinic, those receiving bad conduct discharges were significantly more likely to have been rated by their childhood counselors as having poor peer adjustment than those with successful service records. These data are important for two reasons: they demonstrate a linkage between peer adjustment and moral behavior, and the relation is a predictive one—childhood failure in peer relations was correlated with bad conduct discharge rates in adulthood. In a more recent study (Roff and Sells 1968), a significant relation was demonstrated between peer acceptance–rejection during middle childhood and delinquency during early adolescence. Among upper lower-class and middle-class children, there was a dramatically higher delinquency rate among children who were not accepted by their peers than among those who were. Among the very lowest social class subjects, however, a different pattern held: both highly accepted and highly rejected boys had higher delinquency rates than those who were moderately accepted by their peers. Examination of individual case records indicated that the nature of the delinquency and the adequacy of the child's personality adjustment differed between the chosen and nonchosen lower class groups. In fact, there is every reason to expect that, among the subjects from the lowest social strata, ultimate adjustment of the peer-accepted delinquent group will be better than the nonaccepted delinquent subjects. Thus far, path analysis techniques have not been applied to Roff's data, so that some question still remains concerning the relative primacy of delinquent activity and the poor peer relations. But the data point to peer interaction as one source of the individual's willingness to live according to accepted social standards. . . .

Anxiety and Emotional Disturbance

Evidence can be found in at least 20 studies to show that a child's general emotional adjustment is related to his popularity (Hartup 1970). Assessment of adjustment in these studies has been accomplished with devices as various as the TAT, on the one hand, and observations of school adjustment, on the other. The data consistently show that, in samples of children who are functioning within the normal range, degree of maladjustment is inversely related to degree of social acceptance. In addition to the studies with normal samples, work with institutionalized populations shows that popularity in disturbed groups is also inversely related to relative degree of maladjustment (Davids and Parenti 1958). Sheer quantity of social participation has not been studied in relation to general personality adjustment, although one suspects that such a relation exists.

Some fifteen years ago, a spate of studies was published concerning the relation between a more specific affective component, anxiety, and social acceptance. In general, there tends to be a low negative correlation between anxiety, as measured by the *Children's Manifest Anxiety Scale*, and sociometric status (e.g., McCandless, Castaneda, and Palermo 1956) although, once again, we know very little about the relation between amount of social participation and anxiety.

The major predictive studies of the relation between childhood peer status and adult emotional adjustment have been completed by Roff (1963) within the context of his follow-up investigations of the adult status of boys who, as children, were seen in child guidance clinics. Within these samples, poor peer relations in childhood have been predictive of both neurotic disturbance and psychotic episodes of a variety of types.

Once again, the evidence relating affective disturbance and peer relations is correlational and invites an interpretation suggesting that rejection leads to anxiety, lowered self-esteem, and hostility, which in turn lead to further rejection. A simple unidirectional interpretation does not seem plausible. However, despite the fact that explanatory hypotheses do not receive clear support from this research, there is no evidence that contradicts the basic hypothesis that peer relations are of pivotal importance in personality development (Roff, Sells, and Golden 1972).

CONCLUSION

Access to age-mates, acceptance by them, and constructive interactions with them are among the necessities of child development. There is a relation between the child's personal–social effectance and his sociability with peers, and opportunities for early peer contact appear to enhance the socialization of aggression. At later stages of development, peer contacts provide essential inputs into sexual socialization, although present research does not clearly establish a relation between peer contact and either moral or intellectual development. Major predictive studies, however, have shown that the adequacy of adjustment to the peer group is a good predictor of adolescent and adult emotional status. Thus, no evidence contradicts the notion that peer relations are crucial to a child's development and much evidence supports this contention.

Recent studies also demonstrate that the sensitive student of child behavior has a variety of avenues available through which to maximize the input of peer relations to the development of the individual child. Ensuring each child the opportunity for productive commerce with peers is not an easy task, but such experiences would seem essential to normal social development.

REFERENCES

Baumrind, D. 1972. Socialization and instrumental competence in young children. In W. W. Hartup, ed., *The Young Child: Reviews of Research*, vol. 2. Washington, D.C.: National Association for the Education of Young Children, pp. 202–224.

Bronson, W. C. 1966. Central orientations: a study of behavior organization from childhood to adolescence. *Child Development* 37:125–155.

Campbell, J. D. and M. R. Yarrow. 1961. Perceptual and behavioral correlates of social effectiveness. *Sociometry* 24:1–20.

Davids, A. and A. N. Parenti. 1958. Time orientation and interpersonal relations of emotionally disturbed and normal children. *Journal of Abnormal and Social Psychology* 57:299–305.

Gold, H. A. 1962. The importance of ideology in sociometric evaluation of leadership. *Group Psychotherapy* 15:224–230.

Hamburg, D. A. and J. Van Lawick-Goodall. 1974. Factors facilitating development of aggressive behavior in chimpanzees and humans. In J. de Wit and W. W. Hartup, eds., *Origins and Determinants of Aggressive Behaviors*, pp. 59–86. The Hague: Mouton.

Harlow, H. F. 1969. Agemate or peer affectional system. In D. S. Lehrman, R. A. Hinde, and E. Shaw, eds., *Advances in the Study of Behavior*, 2:333–383. New York; Academic Press.

Hartup, W. W. 1970. Peer interaction and social organization. In P. H. Mussen, ed., *Carmichael's Manual of Child Psychology,* 2:361–456. New York: Wiley.

—— 1974. Aggression in childhood: developmental perspectives. *American Psychologist* 29:336–341.

Heathers, G. 1955. Emotional dependence and independence in nursery school play. *Journal of Genetic Psychology* 87:37–57.

Hetherington, M. and J. Deur. 1972. The effects of father absence on child development. In W. W. Hartup, ed., *The Young Child: Reviews of Research*, 2:303–319. Washington, D.C.: National Association for the Education of Young Children.

Irwin, D. M. 1967. Peer acceptance related to the young child's concept of justice. B. A. thesis, University of Minnesota.

Jay, P., ed. 1968. *Primates: Studies in Adaptation and Variability*. New York: Holt, Rinehart, & Winston.

Keasey, C. B. 1971. Social participation as a factor in the moral development of preadolescents. *Developmental Psychology* 5:216–220.

Kinsey, A. C., W. B. Pomeroy, and C. E. Martin. 1948. *Sexual Behavior in the Human Male*. Philadelphia: W. B. Saunders.

Klaus, R. A. 1958. Interrelationships of attributes that accepted and rejected children ascribe to their peers. Ph.D. dissertation, George Peabody College for Teachers.

Kohlberg, L. 1958. The development of modes of moral thinking and choice in the years ten to sixteen. Ph.D. dissertation, University of Chicago.

Lindzey, G. 1967. Some remarks concerning incest, the incest taboo, and psychoanalytic theory. *American Psychologist* 22:1051–1059.

Lynn, D. B. 1969. *Parental and Sex Role Identification: A Theoretical Formulation*. Berkeley: McCutchan.

Maccoby, E. E. and S. S. Feldman. 1972. Mother-attachment and stranger-reactions in the third year of life. *Monographs of the Society for Research in Child Development*, vol. 37, no. 146.

Maccoby, E. E. and J. C. Masters. 1970. Attachment and dependency. In P. H. Mussen, ed., *Carmichael's Manual of Child Psychology*, 2:73–157. New York: Wiley.

McCandless, B. R., A. Castaneda, and D. S. Palermo. 1956. Anxiety in children and social status. *Child Development* 27:385–391.

Patterson, G. R. and J. A. Cobb. 1971. A dyadic analysis of "aggressive" behaviors. In J. P. Hill, ed., *Minnesota Symposia on Child Psychology*, vol. 5. Minneapolis: University of Minnesota Press.

Patterson, G. R., R. A. Littman, and W. Bricker. 1967. Assertive behavior in children: a step toward a theory of aggression. *Monographs of the Society for Research in Child Development* vol. 32, no. 113.

Piaget, J. 1932. *The Moral Judgment of the Child*. Glencoe, Ill.: Free Press.

Porteus, B. D. and R. C. Johnson. 1965. Children's responses to two measures of conscience development and their relation to sociometric nomination. *Child Development* 36:703–711.

Roff, M. 1961. Childhood social interactions and young adult bad conduct. *Journal of Abnormal and Social Psychology* 63:333–337.

—— 1963. Childhood social interaction and young adult psychosis. *Journal of Clinical Psychology* 19:152–157.

Roff, M. and S. B. Sells. 1968. Juvenile delinquency in relation to peer acceptance-rejection and socioeconomic status. *Psychology in the Schools* 5:3–18.

Roff, M., S. B. Sells, and M. M. Golden. 1972. *Social Adjustment and Personality Development in Children*. Minneapolis: University of Minnesota Press.

Rothbart, M. K. and E. E. Maccoby. 1966. Parents' differential reactions to sons and daughters. *Journal of Personality and Social Psychology* 4:237–243.

Scarr, S. 1969. Social introversion-extraversion as a heritable response. *Child Development* 40:823–832.

Twelve

The Aftermath of Divorce

E. Mavis Hetherington, Martha Cox, and Roger Cox

The number of children growing up in divorced families has increased dramatically in the past few decades. The actual divorce itself is a critical event that affects the child, the parents, and the interaction of the family members with one another. In this article, Hetherington, Cox, and Cox report on the findings of their studies of recently divorced families. Their findings suggest that divorce may have its most dramatic effect on the mother-son relationship.

DIVORCE CAN BE VIEWED as a critical event that affects the entire family system and the functioning and interactions of members within that system. To get a true picture of the impact of divorce, its effects on the divorced parents and on the children must be examined.

The findings reported here are part of a two-year longitudinal study of the impact of divorce on family functioning and children's development. The first goal of the larger study was to examine family responses to the crisis of divorce and then examine patterns of family reorganization over the two-year period following divorce. It was assumed that the family system would go through a period of disorganization immediately after the divorce, followed by recovery, reorganization, and eventual attainment of a new pattern of equilibrium. The second goal was to examine the characteristics of family members which contributed to variations in family processes. The third goal was to examine the effects of variations in family interactions and structure on children's development.

In this selection, we will focus on . . . factors related to alterations in parent-child interactions in the two years following divorce.

148

METHOD

Subjects

The original sample was composed of 72 white, middle-class children (36 boys, 36 girls) and their divorced parents from homes in which custody had been granted to the mother, and the same number of children and parents from intact homes. The mean ages of the divorced mothers and fathers and the mothers and fathers from intact homes were 27.2, 29.6, 27.4, and 30.1 respectively. All parents were high school graduates and the large majority of parents had some college education or advanced training beyond high school. Divorced parents were identified and contacted through court records and lawyers. Only families with a child attending nursery school (who served as the target child) were included in the study. The intact families were selected on the basis of having a child of the same sex, age, and birth order in the same nursery school as the child from a divorced family. In addition, an attempt was made to match parents on age, education, and length of marriage. Only first- and second-born children were included in the study.

The final sample consisted of 24 families in each of four groups (intact families with girls, intact families with boys, divorced families with girls, divorced families with boys)—a total of 96 families for which complete data were available. Sample attrition was largely due to remarriage in the divorced families (19 men, 10 women); separation or divorce in the intact sample (five families); relocation of a family or parent; and lack of cooperation by schools, which made important measures of the children unavailable. Also, eight families no longer wished to participate in the study. Because one of the interests of the investigation was to determine how mothers and children functioned in father-absent homes and how their functioning might be related to deviant or nondeviant behavior in children, families with stepparents were excluded from this study but remained in a stepparent study. In the analyses presented here, six families were randomly dropped from groups to maintain equal sizes of groups.

When a reduction in sample size from 144 families to 96 families occurs, bias in the sample immediately becomes a concern. On demographic characteristics such as age, religion, education, income, occupation, family size, and maternal employment, there were no differences between those subjects who dropped out or were excluded from the sample and those who remained. When a family was no longer included in the study, a comparative analysis was done of its interaction patterns and those of the continuing families. Some differences in these groups will be noted subsequently. In general, there were few differences in parent-child interactions in families who did or did not remain in the study. However, there were some differences in the characteristics of parents who remarried and how they viewed themselves and their lives.

Procedure

The study used a multimethod, multimeasure approach to the investigation of family interaction. The measures used included interviews with, and structured diary records of, the parents, observations of the parents and child interacting in the laboratory and home, behavior checklists of child behavior, parent rating of child behavior, and a battery of personality scales administered to the parents. In addition, observations of the child were conducted in the nursery school. Peer nomination, teacher ratings of the child's behavior, and measures of the child's sex-role typing, cognitive performance, and social development also were obtained. The parents and children were administered these measures two months, one year, and two years after divorce.

Parent Interviews. Parents were interviewed separately on a structured parent interview schedule designed to assess discipline practices and the parent-child relationship; support systems outside the family household system; social, emotional, and heterosexual relationships; quality of the relationship with the spouse; economic stress; family disorganization; satisfaction and happiness; and attitudes toward themselves. The interviews were tape-recorded. . . .

Structured Diary Record. Each parent was asked to complete a structured diary record for three days (one weekday, Saturday, and Sunday). Fathers were asked to include at least one day when they were with their children. The diary record form was divided into half-hour units and contained a checklist of activities, situations, people, and five 7-point bipolar mood rating scales. The dimensions on the mood rating scales included: (1) anxious—relaxed; (2) hostile, angry—friendly loving; (3) unhappy, depressed—happy; (4) helpless—competent, in control; and (5) unloved, rejected—loved.

Each 30-minute unit was subdivided into three 10-minute units. If very different events had occurred in a 30-minute period, the subject was encouraged to record these separately and sequentially. For example, if a father had a fight with his boss and received a phone call from his girl friend in the same half hour, these were recorded sequentially in separate columns. Parents were instructed to check off what they were doing, where they were located, who they were with, and how they were feeling on the mood scales in each 30-minute unit from the time they rose in the morning until they went to sleep at night. The record sheet also left space for any additional comments parents cared to make.

Although parents were encouraged to record at the end of each 30-minute period, because of the situation in which they found themselves, this was sometimes impossible. Any retrospective recording ws noted, and the time the entry was made was also recorded. In the first session, a series of standardized scales dealing with affect, stress, and guilt had been included in the battery of parent measures; however, since the diary mood rating scales were found to be better predictors of behavior than these more time-consuming tests, the standardized scales were subsequently dropped from the study.

Parent-Child Laboratory Interaction. Parents were observed separately interacting with their children in the laboratory in half-hour free play situations and half-hour structured situations involving puzzles, block building, bead stringing, and sorting tasks. The interaction sessions with each parent were scheduled on different days, separated by a period of about a month. One-half of the children interacted with the mother first, and one-half with the father first. All sessions were videotaped to permit multiple coding of behavior. . . .

Checklist of Child Behavior. Although at least three hours of observations of the parent and child interacting in the home situation were collected at three different times, this was not a sufficient time period to obtain an adequate sample of the child's behavior in which we were interested and which occurred relatively infrequently. Parents were given a behavior checklist and a recording form divided into half-hour units, and were asked to record whether a given child behavior had occurred in a particular half-hour period. Three hours of recording were available for fathers, but twenty-four hours were available for mothers. Given behaviors included both acts regarded by parents as noxious, such as yelling, crying, whining, destructiveness, and noncompliance, and those regarded as desirable, such as helping, sharing, cooperative activities, compliance, sustained play, or independent activities.

Parent Rating Scales of Child Behavior. A parent rating scale of child behavior was constructed and standardized on a group of 100 mothers and fathers. Items used in previous observation questionnaires and rating scales, or items which seemed relevant to the interests of this study, were included in an initial pool of 96 items. Parents were asked to rate their children on these items using a 5-point scale, with 1 being never occurs, occurs less often than in most children, and 5 being frequently occurs, occurs more often than in most children. Items which correlated with each other, seemed conceptually related, or had been found to load on the same factor in previous studies, were clustered in seven scales containing a total of 49 items. Only items which correlated with the total score in the scales were retained. Items were phrased to describe very specific behavior, as many of these items were also used on the Checklist of Child Behavior previously described. The seven scales were aggression, inhibition, distractibility, task orientation, prosocial behavior, habit disturbance, and self-control. Divorced parents were asked to rate each item on the basis of the child's current behavior.

Data Analysis. Repeated measure manovas involving test session (two months, one year, two years), sex of child, sex of parent, and family composition (divorced versus intact) were performed for each measure, interview, and laboratory interaction task. Repeated measure manovas were also performed on the mood ratings and the amount of time spent in various activities reported in the structured diary records, on the checklist, and in the rating scales. . . . Correlational analyses of all variables within and across subgroups were also performed. In addition, multiple regression and cross-lagged panel correlations and structural equations were calculated for selected parent and child variables in an attempt to identify functional and causal relationships contributing to changes in the behavior of family members across time.

RESULTS

The results of the study will not be presented separately for each procedure used. Instead, the combined findings of the different procedures will be used to discuss alterations in life-style, stresses, and coping by family members and family relations, and how these factors changed in the two years following divorce.

The interaction patterns between divorced parents and children differed significantly from those of intact families on many variables studied in the interview and on many of the parallel measures in the structured interaction situation. On these measures the differences were greatest during the first year; a process of reequilibration seemed to be taking place by the end of the second year, particularly in mother-child relationships. However, even at the end of the second year, parent-child relations in divorced and intact families still differed on many dimensions. Although there were still many stresses in the parent-child interactions of divorced parents after two years, it is noteworthy that almost one-fourth of the fathers and one-half of the mothers reported that their relationships with their children had improved over those during the marriage when parental conflict and tensions had detrimental effects.

Some of the findings for fathers must be interpreted in view of the fact that divorced fathers became increasingly less available to their children and ex-spouses over the course of the two years. Although at two months divorced fathers were having almost as many face-to-face interactions with their children as fathers in intact homes— who were often highly unavailable to their children (Blanchard and Biller 1971)—these interactions declined rapidly. At two months, about one-fourt of the divorced parents reported that fathers, in their eagerness to maximize visitation rights and maintain contact with their children, were having even more face-to-face contact with their children than they had before the divorce. This contact was motivated by a variety of factors. Sometimes it was based on the father's deep attachment to the child or continuing attachment to the wife; sometimes it was based on feelings of duty or attempts to assuage guilt; often it was an attempt to maintain a sense of continuity in the father's life. Unfortunately, it was often at least partly motivated by a desire to annoy, compete with, or retaliate against the spouse. By two years after the divorce, 19 divorced fathers saw their children once a week or more, 14 fathers saw them every two weeks, 7 every three weeks, and 8 once a month or less.

Results of the diary record, interview findings, and laboratory observations relating to parent-child interactions will be presented in a simplified fashion and, when possible, presented together. The patterns of parent-child interaction showed considerable congruence across these measures.

Divorced parents made fewer maturity demands, communicated less well, tended to be less affectionate, and showed marked inconsistency in discipline and control of their children in comparison to married parents. Poor parenting was most apparent when divorced parents, particularly divorced mothers, interacted with their sons. Divorced parents communicated less, were less consistent, and used more negative sanctions with sons than with daughters. Additionally, in the laboratory situation divorced mothers exhibited fewer positive behaviors (such as positive sanctions and affiliations) and more negative behaviors (such as negative commands, negative sanctions, and opposition to children's requests) with sons than with daughters. Sons of divorced parents seemed to have a difficult time, and this may partly explain why—as we shall see shortly—the adverse effects of divorce are more severe and enduring for boys than for girls.

Fortunately, parents learned to adapt to problem situations, and by two years after divorce the parenting practices of divorced mothers had improved. Poor parenting seemed most marked, particularly for divorced mothers, one year after divorce, which appeared to be a peak of stress in parent-child relations. Two years after divorce mothers were demanding more autonomous, mature behavior of their children, communicated better, and used more explanations and reasoning. They were more nurturant and consistent, and were better able to control their children than before. A similar pattern occurred for divorced fathers in maturity demands, communication, and consistency, but they became less nurturant and more detached from their children with time; in the laboratory and home observations, divorced fathers ignored their children more and showed less affection.

The interviews and observations showed that the lack of control divorced parents had over their children was associated with very different patterns of relating to children by mothers and fathers. The divorced mother tried to control her child by being more restrictive and giving more commands which the child ignored or resisted. The divorced father wanted his contacts with his child to be as happy as possible. He began by being extremely permissive and indulgent with his child and becoming increasingly restrictive over the two-year period, although he was never as restrictive as fathers in intact homes. The divorced mother used more negative sanctions than the divorced father or than parents in intact families. However, by the second year the divorced mother's use of negative sanctions declined as the divorced father's increased. In a parallel fashion, the divorced mother's use of positive sanctions increased after the first year as the divorced father's decreased. The "every day is Christmas" behavior of the divorced father declined with time. The divorced mother decreased her futile attempts at authoritarian control and became more effective in dealing with her child over the two-year period.

The lack of control divorced parents had over their children, particularly one year after divorce, was apparent in both home and laboratory observations. The observed frequency of children's compliance with parents' regulations, commands, or requests could be regarded as a measure of either parental control or resistant child behavior. A clearer understanding of functional relationships in parent-child interaction may be obtained by examining the effectiveness of various types of parental responses in leading to compliance by children and parents' responses to children following compliance or noncompliance. It can be seen in table 12.1 that boys are less compliant than girls and that fathers are more effective than mothers in obtaining compliance from children in both divorced and intact families. This may be at least partly based on the fact that mothers gave over twice as many commands as fathers, and divorced mothers gave significantly more commands than divorced fathers or parents in intact families.

The curvilinear effect—with the least effectiveness of any type of parental behavior at one year and a marked increased in control of the child by two years—is again apparent, although divorced mothers and fathers never gained as much control as their married counterparts. Because developmental psychologists have traditionally regarded reasoning and explanation as the font of good discipline from which all virtues flow, the results relating to types of parental demands were unexpected. Negative commands were less effective than positive commands and, somewhat surprisingly, in the two-month and one-year groups, reasoning and explanation were less effective than

Table 12.1. Compliance with Positive and Negative Parental Commands and Parental Reasoning and Explanation

| | Intact | | | | Divorced | | | |
| | Girl | | Boy | | Girl | | Boy | |
	Father	Mother	Father	Mother	Father	Mother	Father	Mother
Percentage of compliance with positive parental commands								
Two months	60.2	54.6	51.3	42.6	51.3	40.6	39.9	29.3
One year	63.4	56.7	54.9	44.8	43.9	31.8	32.6	21.5
Two years	64.5	59.3	57.7	45.3	52.1	44.2	43.7	37.1
Percentage of compliance with negative parental commands								
Two months	55.7	49.3	47.5	36.4	47.0	34.8	35.6	23.4
One year	59.2	51.5	50.3	38.8	39.1	27.2	28.3	17.2
Two years	60.5	54.6	53.6	39.0	49.9	39.7	39.7	31.8
Percentage of compliance with parental reasoning and explanation								
Two months	49.1	43.3	41.0	31.1	41.3	29.2	29.6	18.4
One year	55.4	48.0	46.2	34.5	26.3	23.1	24.5	14.1
Two year	62.3	58.1	58.1	47.6	50.3	42.5	41.4	36.9

either positive or negative commands. By the last test session the effectiveness of reasoning and explanation significantly increased over the previous sessions. Two things were noteworthy about the pattern of change in reasoning.

First, it should be remembered that the average age of the subjects was two years older at the final session. The mean age of children at the two-month session was 3.92 years; at the one-year session, 4.79 years; and at the final session, 5.81 years. It may be that as children became more cognitively and linguistically mature reasoning and explanation were more effective because children could better understand and had longer attention spans. It may also be that internalization and role-taking were increasing and explanations involving appeals to the rights and feelings of others became more effective. . . .

Second, two years following divorce reasoning was superior to negative parental commands in obtaining compliance from boys (with the exception of sons interacting with their divorced fathers) but not from girls. Why should reasoning be relatively more effective in gaining compliance from boys? Martin (1974) in his recent review of research on parent-child interactions suggests that coercive parental responses are more likely to be related to oversocialization and inhibition in girls, and to aggression in boys. It may be that the greater aggressiveness frequently observed in preschool boys and the greater assertiveness in the culturally proscribed male role necessitate the use of reasoning and explanation to develop the cognitive mediators necessary for self-control in boys. Some support for this idea was found in the greater number of, significantly larger, and more consistent correlations for boys than for girls between the communication scale of the parent interview, frequency of observed parental reasoning and explanation, and parents' ratings of children's prosocial behavior, self-control, and aggression. A similar pattern of correlations was obtained between these parental measures and the frequency of negative and positive behavior on the behavior checklist. In contrast, high use of negative commands was positively related to aggression in boys, but not in girls. Although reasoning and explanation are not clearly superior to other commands in gaining short-term compliance, these methods are more effective in the long-term development of self-control, inhibition of aggression, and prosocial behavior in boys.

We can extend our analysis of compliance one step further and examine how parents respond to compliance or noncompliance by children. Developmental psychologists and behavior modifiers have emphasized the role of contingent reinforcement in effective parenting. Parental responses to compliance are presented in table 12.2, and parental responses to noncompliance are presented in table 12.3; the most frequently occurring responses are included in these tables. Only the most significant effects will be noted.

First, it can be seen that children received positive reinforcement in fewer than one-half of the times they complied—not a very lavish reinforcement schedule for good behavior. Second, boys who complied received less positive reinforcement; more commands, both positive and negative; and more negative sanctions (such as, "You didn't do that very fast" or "You'd better shape up if you know what's good for you") than girls. Boys were not as appropriately reinforced for compliance as girls. This seemed to be the case particularly for divorced mothers and sons across all ages, although divorced mothers became significantly more appropriate in responding to compliance by children from one year to two years after divorce. In contrast, divorced fathers became less reinforcing and attentive to children's positive behaviors in this period.

Table 12.2. Parents' Consequent Behaviors toward Compliance

	Intact				Divorced			
	Girl		Boy		Girl		Boy	
	Father	Mother	Father	Mother	Father	Mother	Father	Mother
Percentage of positive sanctions (affiliate, encourage)								
Two months	39.0	51.1	34.6	49.8	46.6	37.0	44.2	31.8
One year	42.6	49.7	37.2	45.6	47.4	32.5	42.4	28.8
Two years	44.4	49.4	39.8	48.1	36.8	41.6	34.7	37.3
Percentage of ignoring or no response								
Two months	21.4	15.8	18.7	16.3	19.2	28.9	17.6	27.9
One year	23.9	16.2	20.9	17.8	20.4	30.0	17.1	30.3
Two years	22.6	14.9	19.5	16.4	30.2	19.8	25.1	20.0
Percentage of positive commands								
Two months	11.6	16.2	15.1	20.3	6.2	18.8	10.0	20.4
One year	12.3	15.3	14.9	18.6	6.6	20.7	8.3	21.1
Two years	10.9	14.7	15.5	19.9	8.3	17.0	13.5	20.0
Percentage of negative commands or negative sanctions								
Two months	8.6	4.3	10.5	7.6	2.4	8.2	5.0	13.4
One year	6.7	5.1	9.7	6.6	4.2	10.3	8.1	14.9
Two years	6.9	4.8	9.9	6.9	7.6	5.6	9.3	10.5

How did parents respond when children failed to obey their commands? In most cases, they gave another command, sometimes using negative sanctions. Parents in intact families, especially mothers, were also likely to deal with noncompliance by reasoning with children. Sometimes parents, notably fathers, intruded physically by moving the children or surrounding objects. There was much less ignoring of noncompliance than of compliance, especially by fathers in intact families. If parental ignoring responses are examined, it is clear that one way divorced parents coped with noncompliance was by pretending it did not happen. The chains of noncompliance by children, followed by ignoring, were of longer duration in divorced families than in intact families, especially in the interactions of divorced mothers and their sons.

Divorced mothers dramatically increased their use of reasoning and explanation in response to noncompliance in the second year after divorce, while divorced fathers became less communicative and more negative in their responses.

Table 12.3. Parents' Consequent Behaviors toward Noncompliance

	Intact				Divorced			
	Girl		Boy		Girl		Boy	
	Father	Mother	Father	Mother	Father	Mother	Father	Mother
Percentage of positive commands								
Two months	26.3	27.9	24.2	29.3	23.6	20.6	22.0	23.9
One year	24.8	23.6	25.7	27.5	20.5	16.9	20.0	15.3
Two years	25.1	26.8	26.0	27.9	19.1	23.0	18.0	20.6
Percentage of reasoning, explantion, and encouraging positive questions								
Two months	20.1	25.6	17.9	22.7	18.9	20.1	15.0	16.5
One year	22.7	25.3	20.1	23.2	18.1	16.3	16.9	11.9
Two years	26.8	30.4	24.3	28.5	14.0	27.4	13.2	22.7
Percentage of negative commands, negative sanctions, and negative questions								
Two months	18.5	14.0	20.0	18.3	9.0	24.0	12.0	26.2
One year	20.4	15.7	22.5	17.0	9.9	26.8	13.1	28.1
Two years	22.2	17.5	26.8	17.5	19.8	20.2	22.2	21.7
Percentage of ignoring or no response								
Two months	5.3	8.1	4.1	9.9	13.5	15.8	14.3	16.0
One year	4.8	6.7	3.9	7.3	16.7	17.9	18.1	19.3
Two years	4.0	4.9	3.0	5.8	18.1	11.6	21.6	10.2
Percentage of physical intrusion								
Two months	10.8	6.9	9.5	5.8	11.2	8.3	9.3	8.0
One year	8.6	5.7	8.7	5.4	9.1	8.2	8.4	7.5
Two years	7.2	5.8	6.9	5.0	8.1	6.0	7.3	5.9

After reviewing the interview and observational findings, one might be prone to state that disruptions in children's behavior after divorce are attributable to emotional disturbance in the divorced parents and poor parenting, especially by mothers of boys. However, before we point a condemning finger at these parents, especially the divorced mothers who face the day-to-day problems of childrearing, let us look at the children involved. The findings on the behavior checklist, recording the occurrence of children's positive and negative behaviors in the home in 30-minute units, showed not only that children of divorced parents exhibited more negative behavior than children of intact families, but also that these behaviors were most marked in boys and had largely disappeared in girls two years after divorce. Such behaviors were also significantly declining in boys.

Children exhibited more negative behavior with their mothers than with their fathers; this was especially true with sons of divorced parents.

These checklist results were corroborated by the home and laboratory observations, and by parent ratings of children's behavior. Divorced mothers may have given their children a difficult time, but mothers, especially divorced mothers, got rough treatment from their children. As previously remarked, children were more likely to exhibit oppositional behavior to mothers and comply with fathers. The children made negative complaining demands of the mother more frequently. Boys were more oppositional and aggressive; girls were more whining, complaining, and compliant. Children of divorced parents showed an increase in dependency over time, and exhibited less sustained play than children of intact families. The divorced mother was harassed by their children, especially her sons. In comparison with fathers and with mothers in intact families, children of the divorced mother did not obey, affiliate, or attend to her in the first year after divorce. They nagged and whined, made more dependency demands, and were more likely to ignore her. Aggression of sons of divorced mothers peaked at one year, then dropped significantly, but was still higher at two years than aggression of sons in intact families. Some divorced mothers described their relationships with their children one year after divorce as "declared war," "a struggle for survival," "the old water torture," or "getting bitten to death by ducks." One year following divorce seemed to be the period of maximum negative behaviors for children, as it was for the divorced parents themselves. Great improvement occurred by two years, although negative behaviors were more sustained in boys than in girls. The second year appeared to be a period of marked recovery and constructive adaptation for divorced mothers and children.

Who is doing what to whom? It has been proposed—most recently by Patterson in a paper entitled "Mothers: The Unacknowledged Victims" (1976)—that the maternal role is not a very rewarding or satisfying one. Patterson demonstrates that the maternal role, particularly with mothers of problem children, demands high rates of responding with very low levels of positive reinforcement for the mothers. He assumes that mothers and their aggressive children get involved in a vicious circle of coercion. The mother's lack of management skills accelerates the child's aversive behavior of which the mother is the main instigator and for which she is the main target.

This is reciprocated by increased coercion in the mother's parenting behavior, and feelings of helplessness, depression, anger, and self-doubt. In his study, Patterson shows that decreases in the noxious behaviors of aggressive children through treatment procedures aimed at improving parenting skills are associated with decreases of maternal scores on a number of clinical scales on the Minnesota Multiphasic Personality Inventory (MMPI), with a decrease in anxiety on the Taylor Manifest Anxiety Scale, and with improvement on several other measures of maternal adjustment.

Patterson's model may be particularly applicable to divorced mothers and their children in our study. High synchronous correlations between reported and observed poor parenting in divorced mothers and between reported and observed negative behavior in children occurred at each time period. The greater use of poor maternal parenting practices and higher frequency of undesirable behaviors in children from divorced families, even in the first sessions with mothers and sons, suggests that the coercive cycle was already underway when we first encountered our families two months after divorce. Stresses and conflicts preceding or accompanying divorce might have initiated the cycle. High rates and durations of negative exchanges between divorced mothers and their sons were apparent throughout the study. Sequence analyses of the home and laboratory observations showed that divorced mothers of boys were not only more likely than other parents to trigger noxious behavior, but also that they were less able to control or terminate this behavior once it occurred.

We attempted to use cross-lagged panel correlations between selected parent and child measures at the three time periods to identify causal effects in these interactions. Panel correlations were problematic with our study, which involved a relatively small sample size. Kenney (1975), in his review of cross-lagged panel correlations, stated that it is difficult to obtain significant results with Ns under 75. In our study, if we analyzed divorced and intact families separately, but pooled boys and girls, we had only 48 families in a group. Because the family dynamics differed somewhat in families with boys and girls, especially for divorced families, it seemed conceptually unsound to combine sexes, but then we were left with a meager 24 families per group. In spite of these difficulties, we did obtain some findings of interest on the panel correlations.

There were many significant synchronous correlations between parent and child behavior. Poor parenting practices and coercive behavior in parents correlated with undesirable and coercive behavior in children; this was particularly true for divorced mothers and their sons. This suggests that the coercive cycle was already underway. Causal direction for poor parenting practices and noxious child behavior could not be identified consistently by the panel correlations. However, the observational measures and child checklist measures—but not the interview and rating measures—indicated that poor parenting by divorced mothers at two months after divorce caused problem behaviors in children at one year. These effects were similar but not significant between one year and two years.

A striking finding was that divorced mothers' self-esteem, feelings of parental competence as measured by the interview, state anxiety as measured by the Spielberger State-Trait Anxiety Scale, and mood ratings of competence, depression, and anxiety on the structured diary record not only showed significant synchronous correlation with ratings of children's aggression and checklist frequency of noxious behaviors, but also yielded significant cross-lagged panel correlations, suggesting that the behavior of the children—particularly of the sons—was causing the emotional responses of the mother. The findings were similar but less consistent for mothers in intact families. Mothers from divorced and intact families showed more state and trait anxiety, feelings of external control and incompetence, and depression than fathers. This suggests that the feminine maternal role is not as gratifying as the masculine paternal role, regardless of whether the family is intact or divorced. The more marked findings in divorced mothers seemed in accord with Patterson's view that mothers of problem children are trapped in a coercive cycle that leads to debilitating attitudes toward themselves, adverse emotional responses, and feelings of helplessness.

In Patterson's study and others comparing parents of problem and nonproblem children, fathers were found to be much less affected by problem children than were mothers. Fathers, particularly divorced fathers, spent less time with their children than did mothers, thereby escaping some of the stresses imposed by coercive children and obtaining more gratification in activities outside the family. Fathers seemed less likely to get involved in a coercive vicious cycle because children exhibited less deviant behavior in their presence; furthermore, fathers were better able to control deviant behavior by children once it occurred, as was shown in fathers' ratings of children's behavior, frequencies of behavior on the checklist, and observations in the home and laboratory.

The cross-lagged panel correlations showed a larger proportion of effects going in the direction of fathers causing children's behavior rather than in children causing fathers' behaviors, relative to the number found in mother-child interactions. Children's behavior showed few effects on the state anxiety, mood ratings, or self-esteem of fathers, especially divorced fathers. In addition, in intact families, negative child behaviors at the one- and two-year periods seemed to be partially caused by poor control, low nurturance, and high use of negative sanctions by fathers at the earlier periods.

The 48 divorced fathers involved in this study probably showed more concern about their children and interacted with them more than most divorced fathers. The fact that they were available for study and willing to participate may reflect a more sustained and greater degree of paternal involvement than is customarily found. However, despite this possible bias, the impact of divorced fathers on children declined with time, and was significantly less than that of fathers in intact families. At two months following divorce, the number of significant correlations between paternal characteristics and behavior and child characteristics was about the same as in intact families. However, two years after divorce the divorced fathers clearly had less influence with their children while divorced mothers had more influence. Divorced mothers became increasingly salient relative to divorced fathers in the social, cognitive, and personality development of their children. This decrease was less marked for divorced fathers who maintained a high rate of contact with their children.

It would seem that in the period leading to and following divorce, parents go through many role changes and encounter many problems, and that they would benefit from support in coping with these problems.

In both divorced and intact families, effectiveness in dealing with the child was related to support from the spouse in childrearing and agreement with the spouse in disciplining the child. When support and agreement occurred between divorced couples, the disruption in family functioning appeared to be less extreme, and the restabilizing of family functioning occurred earlier—by the end of the first year.

When divorced parents agreed about childrearing, had positive attitudes toward each other, and were low in conflict, and when the divorced father was emotionally mature—as measured by the Socialization Scale of the California Personality Inventory (Gough 1969) and the Personal Adjustment Scale of the Adjective Checklist (Gough and Heilbrun 1965)—frequent contact between father and child was associated with positive mother-child interactions and positive adjustment of the child. Where there were disagreements and inconsistencies in attitudes toward the child and conflict between the divorced parents, or when the father was poorly adjusted, frequent visitation by the father was associated with poor mother-child functioning and disruptions in the child's behavior. Emotional maturity in the mother was also found to be related to her adequacy in coping with stresses in her new single life and her relations with the child.

Other support systems, such as parents, siblings, close friends (especially other divorced friends or intimate male friends), or a competent housekeeper, also were related to the mother's effectiveness in interacting with the child in divorced, but not in intact, families. However, none of these support systems was as salient as a continued, positive, mutually supportive relationship between the divorced couple and continued involvement of the father with the child. For the divorced father, intimate female friends, married friends, and relatives offered the next greatest support in his relationship with the child.

REFERENCES

Blanchard, R. W. and H. B. Biller. 1971. Father availability and academic performance among third grade boys. *Developmental Psychology* 4:301–305.

Gough, H. G. 1969. *Manual for California Personality Inventory*. Palo Alto, Calif.: Consulting Psychologists Press.

Gough, H. G. and A. B. Heilbrun. 1965. *The Adjective Checklist*. Palo Alto, Calif.: Consulting Psychologists Press.

Kenney, D. A. 1975. A quasi-experimental approach to assessing treatment effects in the monequivalent control group design. *Psychological Bulletin* 82:345–362.

Martin, B. 1973. Parent-child relations. In B. M. Caldwell and H. N. Ricciuti, eds., *Review of Child Development Research*. Chicago: University of Chicago Press.

Patterson, G. 1976. Mothers: the unacknowledged victims. Paper presented at the Society for Research in Child Development meeting, Oakland, Calif., April.

Part III

Adolescence
Mastery and Competencies

Thirteen

Understanding the Young Adolescent
David Elkind

Many of the changes in *social* behavior that occur during early adolescence can be understood as reflecting changes in *cognitive* competencies. In this selection, David Elkind discusses four "troublesome" behaviors commonly observed among young adolescents—pseudostupidity, the imaginary audience, the personal fable, and hypocrisy—in light of the development of *formal operations*, a distinctive type of reasoning that first appears during the early adolescent years.

IN A WAY, this title is somewhat misleading. I once had a professor who asserted that if you *really* understood something you could build it. Unfortunately I am not prepared to provide you with a "do it yourself build a young adolescent kit." What I would like to do is to introduce several ideas that may make some types of annoying and/or perplexing teenage behavior more meaningful and, hopefully, less troublesome.

Before I do that, however, it is necessary to set the stage for these ideas. Not surprisingly, any insights I may have about adolescents are, in large part, borrowed from my mentor, Jean Piaget. However, these insights also have to do with the affective as well as with the cognitive domain. Piaget's work and theory have so often been discussed in connection with thinking that their implications for feelings have been overlooked. The ideas I want to present have an affective component with respect to both adults and young people.

To understand the kind of affective significance I have in mind, an example of the impact of some of Freud's ideas may be helpful. At the time that Freud introduced his theory of infantile sexuality, puritan notions regarding sex still held sway. Infantile sexuality was regarded as evil and as evidence of original sin. Parents were encouraged to take extreme measures to combat such pleasure-seeking activities as masturbation and thumbsucking. If a child masturbated, the parent was instructed to sew the sleeves of the child's pajamas outside of the bed clothes. And if the child was a thumbsucker, parents were to sew the sleeves of the child's pajamas beneath the bed clothes.

Of course, there were always those parents unlucky enough to have a child who both masturbated and sucked his thumb. But those who wrote for parents around the turn of the century were undaunted by such a problem. Showing the ingenuity for which America is famous, they encouraged the parents to sew up the ends of the pajama sleeves so that the pleasure-seeking fingers could never attain their goal! Thanks to Freud, such measures are no longer taken. Today we recognize that these behaviors are normal to most young children who very quickly give them up on their own.

What is apparent in this example is that if we don't understand someone else's behavior, we are likely to attribute the worst possible motives to it. Before we understand that thumbsucking and masturbation were normal developmental characteristics, we assumed them to be the result of evil intentions. The same holds true for many behaviors that young adolescents engage in. Because we don't understand them, we attribute their behavior to bad motives. When we are confronted with such behaviors, we believe they are done purposively to frustrate or infuriate us.

It is at this point that Piaget's work has affective, as well as cognitive, significance. By providing insights into the troubling behaviors of adolescents, Piaget permits us to deal with such behavior more calmly and rationally than we might do otherwise. It enables us to overcome our all too human tendency to attribute the worst of all motives to behavior that we don't understand. Accordingly, I want now to describe some troubling behaviors of young adolescents in the context of the intellectual processes that bring them about.

PSEUDOSTUPIDITY

We are all familiar with Edgar Allen Poe's story of the purloined letter. In that story Poe demonstrates that the obvious is overlooked when we anticipate something to be hidden. We all have a tendency at times to respond to situations at a more complex level than is warranted by the situation. Recall the child who was filling out an application and asked his father, "What does 'sex' mean?" After some embarrassed hesitation, the father went into a detailed explanation of the birds and the bees. At the end the child said, "That was very nice but I still don't know what 'sex' means. Do I put a check by the 'M' or the 'F'?"

While the tendency to interpret situations more complexly than is warranted happens to all of us at times, it is much more common to young adolescents. The obvious often seems to elude them. In trying to find a sock or a shoe or a book, they ignore the obvious places and look into the esoteric ones. Simple decisions as to what dress or slacks to wear are overcomplicated by the inclusion of extraneous concerns such as why and by whom the clothes were bought in the first place. In school, young people often approach subjects at a much too complex level and fail, not because the tasks are too difficult, but because they are too simple.

Such behavior on the part of young adolescents is what I call *pseudostupidity*. It derives from the newly attained thinking capacities made possible by what Piaget calls formal operations. Formal operations, which appear at about the age of eleven or twelve in most young people, bring about a Copernican change in young people's thinking. They become capable of holding many variables in mind at the same time, of conceiving ideals and contrary-to-fact propositions, and of comprehending metaphor and simile.

But in the young adolescent, these newly attained formal operations are not fully under control. The capacity to conceive many different alternatives is not immediately coupled with the ability to assign priorities and to decide which choices are most appropriate. Consequently, young adolescents often appear stupid because they are, in fact, too bright.

They seek complex, devious motives in the behavior of their siblings and parents for the most innocent occurrences. And even the simplest interpersonal exchanges can be complicated by the young adolescent's overeager intellectualization. I recall telling my son Paul, when he was thirteen, that he had some pizza sauce on his cheek. I pointed to the left side of my face to indicate it was on the left side of his, but he insisted on taking my point of view and kept wiping the right side of his face until I reached over and wiped it off for him. He not only took my point of view, but assumed that I had not taken his.

Again such *contretemps* derive from the young adolescent's newfound cognitive abilities and not from an early adolescent aggression to cretinism.

THE IMAGINARY AUDIENCE

Recently I was having dinner alone at O'Hare Airport. During the course of the meal I happened to drop my knife, which made a horrible clang as it hit the floor. I was sure, at the moment, that everyone else in the restaurant heard the racket and was looking at me thinking, "What a klutz!" In fact, of course, few people heard it and even those who did, did not care. But at the moment I was surrounded by an audience of my own making, *an imaginary audience*.

Everyone has experienced similar moments. But what happens only occasionally in adults is characteristic of the young adolescent because of the formal operations which make it possible for young people to think about other people's thinking. This newfound ability to think about other people's thinking, however, is coupled with an inability to distinguish between what is of interest to others and what is of interest to the self. Since the young adolescent is preoccupied with his or her own self—all the physical and physiological changes going on—he or she assumes that everyone has the same concern. Young people believe that everyone in their vicinity is as preoccupied with their behavior and appearance as they are themselves—they surround themselves with an imaginary audience.

The imaginary audience helps to account for the super-self-consciousness of the young adolescent. When you believe that everyone is watching and evaluating you, you become very self-conscious. In the lunchroom, on the bus going home, standing in front of the class, the young adolescent feels that he or she is at the center of everyone's attention. It is a different sort of self-consciousness than that experienced by children. The child is self-conscious about appearances, about clothes which are too big or the wrong style. But the young adolescent is more concerned about personal qualities, traits, physical features, and abilities which are unique to himself. Fantasies of singing before an audience, of making a touchdown before a cheering crowd, of playing a concerto in a concert hall are common imaginary audience fantasies in which the individual is the center of everyone's attention.

Groups of adolescents are amusing in this regard, for when they come together, each young person is an actor to himself and a spectator to everyone else. Sometimes groups of adolescents contrive to create an audience by loud and provocative behavior. Because they fail to appreciate what is of interest to themselves and what is of interest to others, they cannot understand adult annoyance at their behavior.

In general, imaginary audience behavior tends to decline with age, as young people come to recognize that each individual person has his or her own preoccupations. To be sure, all of us occasionally have imaginary audience reactions, but these are usually short-lived, as was my experience at the airport. Imaginary audience behavior in adults is a relic of early adolescence which all of us carry with us and to which we revert on occasion. But we need to recognize that it is pervasive in young adolescents and that it accounts both for self-consciousness and for their often boorish public behavior.

Before closing this discussion of imaginary audience behavior, it might be well to mention one of its more pathological and disturbing forms, namely, vandalism. In destroying property the vandal imagines the audience's reactions—how the teachers and principal will look and feel when they see windows and furniture broken. Vandalism, which seems so irrational, so incomprehensible, and so senseless, becomes less so when we recognize that it is done with audience reaction in mind. The vandal is angry and wants to ensure that his audience will be angry too. In committing vandalism, the young person has in mind how the audience will react, and it is the reaction of the imagined audience which motivates the distructive behavior.

THE PERSONAL FABLE

There are other actions that young people engage in which are also perplexing. These actions, however, often appear self-destructive rather than injurious to others. The young girl who gets pregnant or the teenager who experiments with drugs causes us to shake our heads in amazement. They know better, we say; they know the facts of life—how women get pregnant and what the dangers are of playing with addictive drugs. Why, we wonder, are they so intent on harming themselves?

I am sure that the Freudians have a number of answers to this question. But we cognitive types have an answer too, an answer that may complement, rather than contradict, Freudian dynamic interpretation. Consider the young person who believes that he or she is always center stage, at the focus of everyone's attention. It is quite natural under these circumstances to feel that you are someone special and above the usual order of things. Other people will grow old and die but not you; other people will get pregnant but not you; other people will be endangered by drugs but not you. This belief that the individual is special and not subject to the natural laws which pertain to others is what I call the *personal fable*. It is a story that we tell ourselves about ourselves but which isn't true.

The personal fable does have adaptive value. It begins in childhood when youngsters fantasize that they are the favored child, and remnants of it are retained in adulthood where our sense of specialty softens the blows of aging and of career and marital stagnation. But while for the child and the adult the personal fable is in the background, for the young adolescent it is front and center. The self is an all-important preoccupation of the young person who has just attained formal operations, and personal fable attitudes appear in many different forms.

One way the personal fable manifests itself is the failure of the young person to distinguish that which is unique to the self from that which is common to mankind. Indeed, young adolescents make a characteristic mistake. They assume that what is common to everyone is unique to themselves. And conversely, they assume that what is unique to themselves is common to everyone. These personal fable confusions result in behaviors which are as familiar as they are annoying to parents and teachers.

One example of this personal fable confusion is provided by the daughter who says to her mother, "Mother, you just don't know how it feels to be in love," or a son who says to his father, "You just don't understand how much I need that metal detector." The young person typically believes that his or her feelings or needs are unique, special, and beyond the realm of understanding by others, particularly adults. The confusion here is between feelings (of love and affection) and needs (for material things) which are common to everyone and those feelings and needs which are unique to the self.

The reverse confusion is also familiar. This occurs when the young adolescent feels that his or her personal preoccupation and concern is shared by everyone. A young man may feel that his nose is too long, that everyone knows his nose is too long, and that everyone, naturally, thinks he is as ugly as he thinks he is. Arguing with him about the fact that his nose is not too long and that he is, in fact, good-looking has little impact. His belief in his ugliness is part of his reality and there is little point in arguing with another person's reality.

In such circumstances, it seems to me, the only helpful thing that we can do is accept the young person's reality while also encouraging him to check his version of reality against that of others. When working with delinquent adolescents, for example, I quickly discovered that arguing with them about their parents did no good at all. On the contrary, when I agreed with them and said, "Yes, your parents do sound pretty rotten. Wonder how you got stuck with those bummers," they often came to their parents' defense. My guess is that if you agreed with a nice-looking young man that his nose *was* too long and that everyone did think he was ugly, you would get an interesting reaction.

The personal fable accounts, in past at least, for a variety of perplexing and troubling behaviors exhibited by the young teenager. It helps account for what appears to be self-destructive behavior but in fact results from a belief that the young person is special and shielded from harm. "It can happen to others, not to me." And the personal fable also accounts for the young adolescent's self-deprecating and self-aggrandizing behavior. In general, personal fable behavior begins to diminish as young people begin to develop friendships in which intimacies are shared. Once young people begin to share their personal feelings and thoughts, they discover that they are less unique and special than they thought and the sense of loneliness they have in being special and apart from everyone else also diminishes.

APPARENT HYPOCRISY

In general, hypocrisy has to do with a discrepancy between one's words and one's deeds. Young adolescents are often quite hypocritical in this sense. A young man of my acquaintance, for example, often carries on at great length, and with considerable eloquence, about his brothers going into his room and taking his things. And he berates his father for not taking stronger measures against the culprits. This same young man feels no compunction, however, about waltzing into his father's study and using the typewriter and calculator located there—not to mention playing rock music on his father's stereo set (reserved for classical records). It would not be surprising for the father to feel that the son was a hypocrite—and I did!

But I also realized that this *apparent hypocrisy* is but another by-product of formal operations that have not been fully elaborated. When an adult shows hypocritical behavior, we assume that he or she has the capacity to relate theory to practice and to see the intimate connection between the two. But in early adolescence, the capacity to formulate general principles of behavior is not immediately linked up with specific examples.

The young adolescent is in much the same position as the preschool child who fails to say "please" or "thank you," despite having been told many times to do so. The rules regulating "please" and "thank you" are general and the child lacks the ability to see the commonality between diverse "please" and "thank you" situations. In the same way, the adolescent is able to conceptualize fairly abstract rules of behavior but lacks the experience to see their relevance to concrete behavior.

To be sure, the personal fable is at work here too. The young person often believes that rules which hold for everyone else fail to hold for him or her. Not surprisingly, adults regard such behavior as self-serving and are upset by it. Again, however, it has to be remembered that such behavior results from intellectual immaturity rather than from defects in moral character.

The apparent hypocrisy of the young adolescent can be observed at the group, as well as at the individual, level. I recall driving out to the airport one Sunday morning when young people were marching in a "Walk for Water." Sponsors agreed to pay a young person a certain amount of money for every mile he or she walked. The money was to go for testing the water of Lake Ontario and for pollution control. As I drove along the route of the march, I was impressed by how many young people were marching and how very well behaved they all were. I was pleased and began to feel that young people today are not as valueless and materialistic as they have sometimes been described.

It would have been better for my peace of mind and for my assessment of young people had I not returned the next day. As I rode along the route I had taken the day before, I was appalled at the litter: McDonald wrappers and soda and beer cans almost obscured the grass and sidewalks. As I watched the teams of city workers cleaning up the mess, I couldn't help but wonder whether the cost of cleaning up didn't amount to more than was collected. Under the circumstances it would have been easy to tag the young people as hypocrites. Weren't they defacing the environment in the name of walking to protect it?

This adult evaluation is not entirely fair to young people. For the early adolescent, expressing an ideal is tantamount to working for and even attaining it. Young people believe that if they can conceive and express high moral principles, then they have in effect attained them and nothing more in a concrete way needs to be done. Indeed, the pragmatic approach of adults—who believe that ideals have to be worked for and that they cannot be attained at once—is regarded as hypocritical by young people. "Don't trust anyone over 30" means don't trust anyone who recognizes the practical difficulties involved in realizing ideals in everyday life.

Here, I believe, is a fundamental cause of the "generation gap." The idealism of the adolescent clashes with the pragmatism of the adult. If we recognize, however, that adolescent idealism is healthy and that pragmatism is too, we can value adolescent idealism without getting upset at their failure to follow through. As young people begin to engage in meaningful work, they come to appreciate the need to expend effort toward attaining ideals, and in so doing, they enter the adult estate. But a certain amount of idealism is healthy in adulthood too. We need to help young people become more pragmatic without, at the same time, making them cynical about ideals and moral principles.

SUMMARY

I have tried to show that many early adolescent behaviors that adults attribute to bad motives derive instead from intellectual immaturity as described by Piaget. The pseudostupidity of young people actually reflects a lack of control over newly attained mental powers. Adolescent self-consciousness, boorishness, and vandalism result from constructing an imaginary audience that monitors their every move and thought. The imaginary audience is a mental construction made possible by the operation of adolescent intelligence. Complementing the imaginary audience is another mental construction, the personal fable, which also gives rise to troublesome behaviors. Young adolescents sometimes behave as if their feelings or thoughts were unique when they are common to everyone, and they sometimes assume that their own personal evaluation of themselves is automatically shared by everyone. Finally, adolescent hypocrisy is only apparent and is most often a failure to distinguish between the expression of an ideal and its pragmatic realization.

Piaget, then, has helped us to shift a whole new set of behaviors from the realm of the "bad" to the realm of "behavior typical for this age group." However, I am *not* saying that because these behaviors are "normal" we should ignore or neglect them. What I am suggesting is that if we understand why adolescents sometimes act dumb or boorish or insensitive or hypocritical, we can deal with it calmly and without a sense of moral outrage. If we recognize that these behaviors reflect intellectual immaturity, we can ourselves be more rational in our reactions to young people.

So Piaget's work does have affective significance, but of a very special kind. By helping us to understand the behavior of young adolescents, it enables us to respond rationally rather than emotionally. Freud hoped that his psychology would change us so that "Where id was, there shall ego be." Piaget's psychology helps us to change so that "Where moral indignation was, there shall rational understanding be."

Fourteen

Intrapsychic versus Cultural Explanations of the "Fear of Success" Motive

Lynn Monahan, Deanna Kuhn, and Phillip Shaver

During the late 1960s, psychologist Matina Horner introduced the concept "fear of success" as an explanation for commonly observed differences in males' and females' performances in competitive, achievement situations. Horner argued that women are more likely than men to feel anxious about achievement because they fear some of the negative consequences of success. In this cleverly designed study, Monahan, Kuhn, and Shaver demonstrate that not only do women feel somewhat ambivalent about success, but that *men* feel ambivalent about *women's* success as well. Negative responses to women's achievement may be more of a cultural, than a psychological, phenomenon.

HORNER (1968, 1970) has offered a widely cited explanation for the failure of many investigators to replicate with female subjects a large body of findings on achievement motivation in males. . . . Two key findings have been at issue: achievement-related fantasies increase under certain arousal conditions, and achievement motivation is positively related to various performance measures (Atkinson and Feather 1966). Neither of these reliable findings for males has been consistently or unambiguously replicated among females. Horner suggested that females may be hampered by a type of achievement-related anxiety uncommon in males. To explore this possibility, she asked female college students to tell a story based on the following cue: After first-term finals, Ann finds herself at the top of her medical school class.

A majority of college women, over 65 percent, portrayed Ann as anxious or guilty or predicted that Ann's success would have negative consequences, such as loss of femininity and social rejection. In a comparable group of college men responding to a similar cue (After first-term finals, John finds himself at the top of his medical school class), less than 10 percent expressed negative themes.

The motive revealed by this projective technique has been labeled "fear of success" by Horner. College-age women, it appears, unlike their male counterparts, fear negative consequences. Moreover, as Horner (1968, 1970) has also shown, women who express fear of success on the projective measure perform worse in a competitive than a noncompetitive situation; whereas most men, and the minority of women who do not show fear of success, perform better under competitive conditions.

These data tell us little, however, about the proper conceptualization of fear of success, its origins, or how it might be eliminated. Horner's reliance on the McClelland and Atkinson approach to motivation implies that she accepts McClelland's (1958) "arguments rooted in accepted principles of learning [supporting] the view, advanced chiefly in psychoanalytic writings, that motives are developed early in childhood and become relatively stable attributes of personality which are highly resistant to change" (Atkinson 1958:598). The present study replicates Horner's design and extends it by completely crossing subject and task factors, a feature absent in Horner's study. This addition to the design, we believe, allows us to distinguish and evaluate two difference interpretations of the fear of success phenomenon. One involves the concept of belief (adherence to cultural stereotypes), and the other is based on a more psychodynamic view of femininity (e.g., Freud 1965).

In Horner's study, the actor in the cue (Ann or John) is of the same sex as the subject, presumably to encourage the subject to identify with the actor and hence reveal his or her own motives. While this projective technique is well accepted in both clinical and research settings, its use in Horner's study makes it impossible to determine whether the task factor or the subject factor is responsible for the experimental effect. In other words, are fear of success responses a function of sex of the subject or sex of the actor in the cue situation?

The present design consists of four groups: (a) females presented with the female cue; (b) females presented with the male cue; (c) males presented with the male cue; and (d) males presented with the female cue. If fear of success responses occur only among females, an "intrapsychic" explanation is suggested: Something about females causes them to have ambivalent or negative attitudes toward achievement. On the other hand, if fear of success responses occur only with the female cue but for both sexes, a "cultural" explanation is suggested: the stereotypes surrounding women's achievements are negative ones, learned and accepted by both sexes.

A second variable critical in understanding the nature and origins of fear of success is age. If the phenomenon represents either an emerging personality trait or knowledge of a cultural stereotype, it should appear early and, most important, increase, rather than decrease, with age. Thus, the present study attempts to replicate the Horner findings among a sample of boys and girls in the sixth through eleventh grades.

METHOD

Subjects

Subjects were sixth- through eleventh-grade students in a middle-class urban school. Ages ranged from 10 to 16 years. For purposes of analysis, they were divided into two age groups, preadolescent (10–13) and early adolescent (14–16). The total N was 120, with 79 subjects in the 10–13 group and 41 in the 14–16 group. There were 52 boys and 68 girls. Twenty-eight boys and 35 girls were given the female cue, and the remaining subjects were given the male cue.

Procedure

The subjects were tested in classroom groups. The female experimenter was introduced to the class, and it was explained that she had a story-telling game for them to play. The class was told that this was not a test, that they would not put their names on the papers turned in to the experimenter, and that their teacher would not see the papers. The aim was to replicate Horner's procedure as closely as possible. Since these subjects were younger than Horner's, however, and had no previous experience with projective tests, the instructions were made more explicit by presenting an example prior to the test cue. The experimenter read the following sentence: "Peter has just learned that he has won the New York State lottery for $5,000." The experimenter then explained that the object of the game was to make up a story starting with that sentence. It was stressed that there was no right or wrong answer; each person should make up his or her own story. Several possible stories, all neutral in content, were then presented and discussed.

Following this group discussion, the test cues were presented. Each subject received a sheet of paper at the top of which was printed either a male or female cue: "After first-term finals, Ann (John) finds herself (himself) at the top of her (his) medical school class." The same "probes" used by Horner appeared below the cue, although the wording was altered slightly to make them simpler and more concrete:

1. Describe Ann (John). What is she like?
2. What is the reaction to the news?
3. What do Ann, and possibly others involved, think after hearing the news?
4. What does Ann want now?
5. What has Ann's life been like up to this point? Tell about the events leading up to this situation.
6. What does the future hold for Ann and those involved with her?

Each subject was given as much time as necessary. When all of the subjects were finished, the experimenter collected the papers. All of the subjects accepted the instructions readily and were quite cooperative. Many wrote lengthy stories, and all of the protocols were adequate for scoring.

Scoring

The scoring procedure was identical to Horner's. If the protocol expressed an overall positive attitude toward the achievement and toward the actor, it was scored as positive. If the protocol expressed any negative attitudes toward the achievement or the actor or specified negative consequences of the achievement, it was scored as negative. The criteria specified by Horner (1970:59) were followed as closely as possible. A subset of 36 protocols were scored by a second rater; percentage agreement in assignment to the positive or negative category was 90 percent. Each negative protocol was further scored as to type of negative imagery, according to the same categories used by Horner: (a) negative consequences because of the success; (b) anticipation of negative consequences because of the success; (c) negative affect because of the success; (d) instrumental activity away from present or future success, including leaving the field for more traditional female work such as nursing, schoolteaching, or social work; (e) any direct expression of conflict about success; (f) denial of effort in attaining the success (also cheating or any other attempt to deny responsibility or reject credit for the success); (g) denial of the situation described by the cue; and (h) bizarre, inappropriate, unrealistic, or nonadaptive responses to the situation described by the cue.

RESULTS

The percentages of subjects whose protocols were assigned to the negative category are given in table 14.1 by condition, by sex, and by age group within sex. The results replicated Horner's: 21 percent of the boys gave negative responses to the John cue, and 51 percent of the girls gave negative responses to the Ann cue. This differences was slightly less extreme than Horner's; her percentages were 65 percent for college women and 10 percent for college men.

The critical comparisons were those across sex, holding cue constant, and across cue, holding sex constant. As some of the com-

Table 14.1. Percentage of Subjects Obtaining Negative Scores

	Actor in the cue	
Subjects	John	Ann
Total male group	21	68
11–13-year-olds	14	75
14–16-year-olds	30	50
Total female group	30	51
11–13-year-olds	29	63
14–16-year-olds	33	27

parisons involved small frequencies, the Fisher exact test was used throughout. For both sexes, the proportion of negative responses to the Ann cue was higher than the proportion of negative responses to the John cue. This difference was highly significant for boys (p = .0006) and just below the traditional significance level for girls (p = .07). Differences across sex showed an interaction: for the John cue, there was no significant difference between sexes in proportion of negative responses ($p < .18$); for the Ann cue, however, boys showed an even higher proportion of negative responses than girls did (p = .08). These results indicate that the sex of the actor presented in the cue, rather than the sex of the subject, was the more critical variable, with a moderate interaction effect in the direction of the boys showing a higher proportion of negative responses to the Ann cue than the girls.

There were no significant differences between age groups in response to the John cue. In response to the Ann cue, both boys and girls showed a decline in negative responses with age, though the decline reached statistical significance only for girls (63 percent versus 27 percent, p = .05).

An examination of types of negative responses by sex and cue suggested an interaction effect: among all of the subjects giving negative responses, girls in response to the Ann cue gave a higher proportion of inappropriate or "nonadaptive" responses. These included denial of the situation, either direct (categories f and g) or by means of leaving the field (category d), and bizarre responses (category h). Some examples are as follows:

> Ann's father was a doctor and her mother was a nurse and she wanted to become like her mother and be a nurse also. So she wanted to go to medical school.

> After Ann went home she got to thinking she does not want to be a nurse or a doctor. . . . She wants to be a teacher. Ann has finally made up her mind. She is not going to be a doctor anymore. She is going to go to college and become a teacher.

> Soon Ann became one of the leading doctors in the world. When she was in France, she met an American man. They both fell in love. Soon they were married. But after they had their first child, Ann turned all her attention to her work. So they got di-

vorced. Ann always was involved in her work. The only people she talked to were fellow doctors and the nurses. Soon she got very ill and died. No one even went to her funeral because she was very mean.

Ann looks like a telephone pole and has purple eyes. Ann is a person who is a mental case which likes to cut up people . . . we worry about the unfortunate people who have to have her as their doctor.

Responses in the other negative categories (a, b, c, and e), in contrast, were characterized simply by an expression of negative affect or negative consequences surrounding the actor; for example:

After hearing the news, Ann realized that she was the only one that deserved it. This proves how conceited our friend is. Everybody thinks she's a bitch. Ann cares only for herself.

The percentage of nonadaptive (categories d, f, g, and h), negative responses on the part of girls responding to the Ann cue was 52 percent. Though the sample was too small to permit conclusive interpretation, it is interesting that all of the older girls gave such responses. Among the other groups, in contrast, percentages of nonadaptive negative responses were 29 percent (boys, Ann cue), 18 percent (boys, John cue), and 24 percent (girls, John cue).

DISCUSSION

Our results replicate a finding that has become widely known and is thought to have profound social implications. In response to a projective cue describing a female's success in a traditionally masculine field, girls in junior high school, senior high school, and college express themes involving considerable negativity and conflict. The present results indicate, however, that interpretation of this finding requires consideration of a complex set of factors.

Males, to an even greater extent than females, responded negatively to the female cue. Neither sex, however, showed a high proportion of negative responses to the male cue. This suggests that negative responses to female achievement, because they occur in both sexes, reflect beliefs that females embarking on a professional career in a traditionally male field anticipate, and to a considerable extent experience, all sorts of difficulties, hardships, and internal and external conflicts. Successful females are often viewed by both sexes as unattractive, immoral, and dissatisfied. In other words, subjects of both sexes are indicating their knowledge of prevailing sex role stereotypes; they believe that women who succeed in a traditionally masculine field encounter a multitude of difficulties. It is not safe, however, to proceed from data concerning beliefs to the psychodynamic inference that females suffer from fear of success motivation. Would females, for example, have reacted negatively to a women's success in a traditionally feminine field such as nursing?[1]

Much of the earlier research on achievement motivation in women has been based on Thematic Apperception Test pictures of males in culturally stereotyped achievement situations. Given our results, it is not surprising that females have expressed achievement-related imagery in response to these cues, or that the resulting achievement motivation scores do not relate consistently to task performance. There is no reason for females' knowledge of the male cultural stereotype to be related to their success on a laboratory task. Similarly, we would not expect males' fear of success responses to a female cue to be related to their performance on a task.

Both Horner's results and the present results, then, can be accounted for in terms of both sexes' belief in conventional sex role stereotypes. There is evidence, however, that something deeper than belief is being tapped. Recall that while both sexes expressed awareness of the negative female stereotype, the types of negative responses given by girls to the female cue were somewhat different. Girls, especially at the older age level, tended to deny the circumstances of the story, foresee bizarre or inappropriate consequences, or suggest events that removed Ann entirely from this evidently conflictful situation. Thus, experiences related to cultural sex role stereotypes may lead to internal conflict, or perhaps awareness of potential conflict, among adolescent girls. An examination of the boys' protocols revealed that many of those expressing a negative reaction to the female cue also had strongly emotional themes. These were primarily sexual in nature and indicated hostility toward successful

---

[1] A recently completed study by Marlaine Katz of Stanford University suggests not.

females; for example:

> She is so overwhelmed she celebrates by letting all the boys lay her as she goes on studying. The future holds for Ann that she will go from whore to prostitute.

Finally, the age trends suggest that negative responses to female achievement decline during adolescence, a trend seemingly inconsistent with Horner's finding of a high proportion of negative responses among college-age women. However, these results may reflect generational, or cohort, rather than longitudinal differences. Horner's subjects were tested five years prior to those in the present study, and during this time women's liberation has received much public attention. Among the older girls in our sample, the proportion of negative responses to the Ann cue decreased significantly, indicating perhaps that they have become aware of the social issues surrounding female achievement and are changing their attitudes accordingly.

REFERENCES

Atkinson, J. W., ed. 1958. *Motives in Fantasy, Action, and Society*. Princeton, N.J.: Van Nostrand.

Freud, S. 1965. Femininity. In S. Freud, ed., *New Introductory Lectures on Psychoanalysis*. New York: Norton.

Horner, M. S. 1968. Sex differences in achievement motivation and performance in competitive and non-competitive situations. Ph.D. dissertation, University of Michigan.

—— 1970. Femininity and successful achievement: a basic inconsistency. In J. Bardwick, E. M. Douvan, M. S. Horner, and D. Gutmann, eds., *Feminine Personality and Conflict*. Belmont Calif.: Brooks-Cole.

McClelland, D. C. 1958. The importance of early learning in the formation of motives. In J. W. Atkinson, ed., *Motives in Fantasy, Action, and Society*. Princeton, N.J.: Van Nostrand.

Adolescence
Identity and the Self

Fifteen

The Problem
of Ego Identity

Erik H. Erikson

No single writer has shaped our ideas about identity development during
adolescence as powerfully as has Erik Erikson. In this selection, Erikson
discusses the development of a *sense of identity*, an important life crisis of the
adolescent years. Erikson's viewpoint helps us to understand that going
through what has come to be called an "identity crisis" may well be a
normal, and indeed necessary, part of growing up.

ADOLESCENCE IS THE LAST and the concluding stage of childhood.
The adolescent process, however, is conclusively complete only when
the individual has subordinated his childhood identifications to a new
kind of identification, achieved in absorbing sociability and in competitive apprenticeship with and among his age-mates. These new
identifications are no longer characterized by the playfulness of childhood and the experimental zest of youth: with dire urgency they force
the young individual into choices and decisions which will, with increasing immediacy, lead to a more final self-definition, to irreversible
role pattern, and thus to commitments "for life." The task to be performed here by the young person and by his society is formidable;
it necessitates, in different individuals and in different societies, great
variations in the duration, in the intensity, and in the ritualization of
adolescence. Societies offer, as individuals require, more or less
sanctioned intermediary periods between childhood and adulthood,
institutionalized *psychosocial moratoria*, during which a lasting pattern of "inner identity" is scheduled for relative completion.

 In postulating a "latency period" which precedes puberty, psychoanalysis has given recognition to some kind of *psychosexual moratorium* in human development—a period of delay which permits the

future mate and parent first to "go to school" (i.e., to undergo whatever schooling is provided for in his technology) and to learn the technical and social rudiments of a work situation. It is not within the confines of the libido theory, however, to give an adequate account of a second period of delay, namely, adolescence. Here the sexually matured individual is more or less retarded in his psychosexual capacity for intimacy and in the psychosocial readiness for parenthood. The period can be viewed as a *psychosocial moratorium* during which the individual through free role experimentation may find a niche in some section of his society, a niche which is firmly defined and yet seems to be uniquely made for him. In finding it the young adult gains an assured sense of inner continuity and social sameness which will bridge what he *was* as a child and what he is *about to become*, and will reconcile his *conception of himself* and his *community's recognition* of him.

If, in the following, we speak of the community's response to the young individual's need to be "recognized" by those around him, we mean something beyond a mere recognition of achievement; for it is of great relevance to the young individual's identity formation that he be responded to, and be given function and status as a person whose gradual growth and transformation make sense to those who begin to make sense to him. . . .

Linguistically as well as psychologically, identity and identification have common roots. Is identity, then, the mere sum of earlier identifications, or is it merely an additional set of identifications?

The limited usefulness of the *mechanism of identification* becomes at once obvious if we consider the fact that none of the identifications of childhood (which in our patients stand out in such morbid elaboration and mutual contradiction) could, if merely added up, result in a functioning personality. True, we usually believe that the task of psychotherapy is the replacement of morbid and excessive identifications by more desirable ones. But as every cure attests, "more desirable" identifications tend to be quietly subordinated to a new, a unique Gestalt which is more than the sum of its parts. The fact is that identification as a mechanism is of limited usefulness. Children, at different stages of their development, identify with those *part aspects* of people by which they themselves are most immediately affected, whether in reality or fantasy. Their identifications with parents, for example, center in certain overvalued and ill-understood body parts, capacities, and role appearances. These part aspects, furthermore, are favored not because of their social acceptability (they often are everything but the parents' most adjusted attributes) but by the nature of infantile fantasy which only gradually gives way to more realistic anticipation of social reality. The final identity, then, as fixed at the end of adolescence is superordinated to any single identification with individuals of the past: it includes all significant identifications, but it also alters them in order to make a unique and a reasonably coherent whole of them. . . .

Identity formation, finally, begins where the usefulness of identification ends. It arises from the selective repudiation and mutual assimilation of childhood identifications, and their absorption in a new configuration, which in turn, is dependent on the process by which a *society* (often through subsocieties) *identifies the young individual*, recognizing him as somebody who had to become the way he is, and who, being the way he is, is taken for granted. The community, often not without some initial mistrust, gives such recognition with a (more or less institutionalized) display of surprise and pleasure in making the acquaintance of a newly emerging individual. For the community, in turn, feels "recognized" by the individual who cares to ask for recognition; it can, by the same token, feel deeply—and vengefully—rejected by the individual who does not seem to care.

While the end of adolescence thus is the stage of an overt identity *crisis*, identity *formation* neither begins nor ends with adolescence: it is a lifelong development largely unconscious to the individual and to his society. Its roots go back all the way to the first self-recognition: in the baby's earliest exchange of smiles there is something of a *self-realization coupled with a mutual recognition.*

The final assembly of all the converging identity elements at the end of childhood (and the abandonment of the divergent ones) appears to be a formidable task: how can a stage as "abnormal" as adolescence be trusted to accomplish it? Here it is not unnecessary to call to mind again that in spite of the similarity of adolescent "symptoms" and episodes to neurotic and psychotic symptoms and episodes, adolescence is not an affliction but a *normative crisis*, i.e., a normal phase of increased conflict characterized by a seeming fluctuation in ego strength, and yet also by a high growth potential. Neurotic and psychotic crises are defined by a certain self-perpetuating propensity, by an increasing waste of defensive energy, and by a deepened psychosocial isolation; while normative crises are relatively more reversible, or, better, traversable, and are characterized by an abundance of available energy which, to be sure, revives dormant anxiety and arouses new conflict, but also supports new and expanded ego functions in the searching and playful engagement of new opportunities and associations. What under prejudiced scrutiny may appear to be the onset of a neurosis is often but an aggravated crisis which might prove to be self-liquidating and, in fact, contributive to the process of identity formation.

It is true, of course, that the adolescent, during the final stage of his identity formation, is apt to suffer more deeply than he ever did before (or ever will again) from a diffusion of roles; and it is also true that such diffusion renders many an adolescent defenseless against the sudden impact of previously latent malignant disturbances. In the meantime, it is important to emphasize that the diffused and vulnerable, aloof and uncommitted, and yet demanding and opinionated personality of the not-too-neurotic adolescent contains many necessary elements of a semideliberate role experimentation of the "I dare you" and "I dare myself" variety. Much of this apparent diffusion thus must be considered *social play* and thus the true genetic successor of childhood play. Similarly, the adolescent's ego development demands and permits playful, if daring, experimentation in fantasy and *introspection*. . . .

Is the sense of identity conscious? At times, of course, it seems only too conscious. For between the double prongs of vital inner need and inexorable outer demand, the still experimenting individual may become the victim of a transitory extreme *identity consciousness* which is the common core of the many forms of "self-consciousness" typical for youth. Where the processes of identity formation are prolonged (a factor which can bring creative gain), such preoccupation with the "self-image" also prevails. We are thus most aware of our identity when we are just about to gain it and when we (with what motion pictures call "a double take") are somewhat surprised to make its acquaintance; or, again, when we are just about to enter a crisis and feel the encroachment of identity diffusion—a syndrome to be described presently.

An increasing sense of identity, on the other hand, is experienced preconsciously as a sense of psychosocial well-being. Its most obvious concomitants are a feeling of being at home in one's body, a sense of "knowing where one is going," and an inner assuredness of anticipated recognition from those who count. Such a sense of identity, however, is never gained nor maintained once and for all.

Like a "good conscience," it is constantly lost and regained, although more lasting and more economical methods of maintenance and restoration are evolved and fortified in late adolescence. . . .

Identity appears as only one concept within a wider conception of the human life cycle which envisages childhood as a *gradual unfolding of the personality through phase-specific psychosocial crises:* I have, on other occasions, expressed this *epigenetic principle* by taking recourse to a diagram which, with its many empty boxes, at intervals may serve as a check on our attempts at detailing psychosocial development. (Such a diagram, however, can be recommended to the serious attention only of those who can take it *and* leave it.) The diagram (figure 15.1), at first, contained only the double-lined boxes along the descending diagonal ı, 1—ıı, 2—ııı, 3—ıv, 4—v, 5—vı, 6—vıı, 7—vııı, 8) and for the sake of initial orientation, the reader is requested to ignore all other entries for the moment. The *diagonal* shows the sequence of psychosocial crises. Each of these boxes is shared by a criterion of relative psychosocial health and the corresponding criterion of relative psychosocial ill-health: in "normal" development, the first must persistently outweigh (although it will never completely do away with) the second. The sequence of stages thus represents a successive development of the component parts of the psychosocial personality. Each part exists in some form (verticals) before the time when it becomes "phase-specific," i.e., when "its" psychosocial crisis is precipitated both by the individual's readiness and by society's pressure. But each component comes to ascendance and finds its more or less lasting solution at the conclusion of "its" stage. It is thus *systematically related* to all the others, and all depend on the proper development at the proper *time* of each; although individual makeup and the nature of society determine the rate of development of each of them, and thus the *ratio* of all of them. It is at the end of adolescence, then, that identity becomes phase-specific (v, 5), i.e., must find a certain integration as a relatively conflict-free psychosocial arrangement—or remain defective or conflict-laden.

With this chart as a blueprint before us, let me state first which aspects of this complex matter will *not* be treated in this paper: for one, we will not be able to make more definitive the now very tentative designation (in *vertical* 5) of the precursors of identity in the infantile ego. Rather, we approach childhood in an untraditional manner, namely, from young adulthood backward—and this with the conviction that early development cannot be understood on its own terms

	1	2	3	4	5	6	7	8
I. INFANCY	Trust vs. Mistrust				Unipolarity vs. Premature Self-Differentiation			
II. EARLY CHILDHOOD		Autonomy vs. Shame, Doubt			Bipolarity vs. Autism			
III. PLAY AGE			Initiative vs. Guilt		Play Identification vs. (oedipal) Fantasy Identities			
IV. SCHOOL AGE				Industry vs. Inferiority	Work Identification vs. Identity Foreclosure			
V. ADOLESCENCE	Time Perspective vs. Time Diffusion	Self-Certainty vs. Identity Consciousness	Role Experimentation vs. Negative Identity	Anticipation of Achievement vs. Work Paralysis	Identity vs. Identity Diffusion	Sexual Identity vs. Bisexual Diffusion	Leadership Polarization vs. Authority Diffusion	Ideological Polarization vs. Diffusion of Ideals
VI. YOUNG ADULT					Solidarity vs. Social Isolation	Intimacy vs. Isolation		
VII. ADULTHOOD							Generativity vs. Self-Absorption	
VIII. MATURE AGE								Integrity vs. Disgust, Despair

Figure 15.1

194

alone, and that the earliest stages of childhood cannot be accounted for without a unified theory of the whole span of preadulthood.

What traditional source of psychoanalytic insight, then, *will* we concern ourselves with? It is: first pathography; in this case the clinical description of *identity diffusion*. Hoping thus to clarify the matter of identity from a more familiar angle, we will then return to the overall aim of beginning to "extract," as Freud put it, "from psychopathology what may be of benefit to normal psychology."

PATHOGRAPHIC: THE CLINICAL PICTURE OF IDENTITY-DIFFUSION

Time of Breakdown

A state of acute identity diffusion usually becomes manifest at a time when the young individual finds himself exposed to a combination of experiences which demand his simultaneous commitment to *physical intimacy* (not by any means always overtly sexual), to decisive *occupational choice*, to energetic *competition*, and to *psychosocial self-definition*. A young college girl, previously overprotected by a conservative mother who is trying to live down a not-so-conservative past, may, on entering college, meet young people of radically different backgrounds, among whom she must choose her friends and her enemies; radically different mores especially in the relationship of the sexes which she must play along with or repudiate; and a commitment to make decisions and choices which will necessitate irreversible competitive involvement or even leadership. Often she finds among very "different" young people a comfortable display of values, manners, and symbols for which one or the other of her parents or grandparents is covertly nostalgic, while overtly despising them. Decisions and choices and, most of all, successes in any direction bring to the fore conflicting identifications and immediately threaten to narrow down the inventory of further tentative choices; and, at the very moment when time is of the essence, every move may establish a binding precedent in psychosocial self-definition, i.e., in the "type" one comes to represent in the types of the age-mates (who seem so terribly eager to type). On the other hand, any marked *avoidance of choices* (i.e., a moratorium by default) leads to a sense of outer isolation. . . .

The Problem of Intimacy

Figure 15.1 shows "Intimacy vs. Isolation" as the core conflict which follows that of "Identity vs. Identity Diffusion." That many of our patients break down at an age which is properly considered more preadult than postadolescent is explained by the fact that often only an attempt to engage in intimate fellowship and competition or in sexual intimacy fully reveals the latent weakness of identity.

True "engagement" with others is the result and the test of firm self-delineation. Where this is still missing, the young individual, when seeking tentative forms of playful intimacy in friendship and competition, in sex play and love, in argument and gossip, is apt to experience a peculiar strain, as if such tentative engagement might turn into an interpersonal fusion amounting to a loss of identity, and requiring, therefore, a tense inner reservation, a caution in commitment. Where a youth does not resolve such strain he may isolate himself and enter, at best, only stereotyped and formalized interpersonal relations; or he may, in repeated hectic attempts and repeated dismal failures, seek intimacy with the most improbable partners. For where an assured sense of identity is missing, even friendships and affairs become desperate attempts at delineating the fuzzy outlines of identity by mutual narcissistic mirroring: to fall in love then often means to fall into one's mirror image, hurting oneself and damaging the mirror. . . .

It must be remembered that the counterpart of intimacy is *distantiation*, i.e., the readiness to repudiate, to ignore, or to destroy those forces and people whose essence seems dangerous to one's own. Intimacy with one set of people and ideas would not be really intimate without an efficient repudiation of another set. Thus, weakness or excess in repudiation is an intrinsic aspect of the inability to gain intimacy because of an incomplete identity: whoever is not sure of his "point of view" cannot repudiate judiciously. . . .

Diffusion of Time Perspective

In extreme instances of delayed and prolonged adolescence, an extreme form of a disturbance in the *experience of time* appears which, in its milder form, belongs to the psychopathology of everyday adolescence. It consists of a sense of great urgency and yet also of a loss of consideration for time as a dimension of living. The young person may feel simultaneously very young, and in fact baby-like, and old beyond rejuvenation. Protests of missed greatness and of a premature and fatal loss of useful potentials are common among our patients as they are among adolescents in cultures which consider such protestations romantic; the implied malignancy, however, consists of a decided disbelief in the possibility that time may bring change, and yet also of a violent fear that it might. This contradiction is often expressed in a general slowing up which makes the patient behave, within the routine of activities (and also of therapy) as if he were moving in molasses. It is hard for him to go to bed and to face the transition into a state of sleep, and it is equally hard for him to get up and face the necessary restitution of wakefulness; it is hard to come to the hour, and hard to leave it. . . .

Diffusion of Industry

Cases of severe identity diffusion regularly also suffer from an acute upset in the sense of workmanship, and this either in the form of an inability to concentrate on required or suggested tasks, or in a self-destructive preoccupation with some one-sided activities, i.e., excessive reading. . . .

The Choice of the Negative Identity

The loss of a sense of identity often is expressed in a scornful and snobbish hostility toward the roles offered as proper and desirable in one's family or immediate community. . . .

On the whole, our patients' conflicts find expression in a more subtle way than the abrogation of personal identity: they rather choose a *negative identity*, i.e., an identity perversely based on all those identifications and roles which, at critical stages of development, had been presented to the individual as most undesirable or dangerous, and yet also as most real.

Such vindictive choices of a negative identity represent, of course, a desperate attempt at regaining some mastery in a situation in which the available positive identity elements cancel each other out. The history of such a choice reveals a set of conditions in which it is easier to derive a sense of identity out of a *total* identification with that which one is *least* supposed to be than to struggle for a feeling of reality in acceptable roles which are unattainable with the patient's inner means. The statement of a young man, "I would rather be quite insecure than a little secure," and that of a young woman, "At least in the gutter I'm a genius," circumscribe the relief following the total choice of a negative identity. . . . Many a late adolescent, if faced with continuing diffusion, would rather *be nobody or somebody bad, or indeed, dead—and this totally, and by free choice—than be not-quite-somebody.*

Sixteen

Disturbance in the Self-Image at Adolescence

Roberta G. Simmons, Florence Rosenberg,
and Morris Rosenberg

It is widely believed that adolescence is a difficult time for the individual. We tend to think of young teenagers as being overly self-conscious, self-critical, and anxious about how others view them. In this selection, Simmons, Rosenberg, and Rosenberg examine this characterization in a study of the self-images of children, young adolescents, and older adolescents. Their findings suggest that, indeed, there may well be a temporary "disturbance" in the self-image that occurs during the early part of adolescence. This tends to support Erikson's view (see selection 15) that adolescence may be a time of major changes in the individual's sense of self.

AMONG THE MOST widely accepted ideas in the behavioral sciences is the theory that adolescence is a period of disturbance for the child's self-image. Hall (1904) originally characterized the age as one of "storm and stress." Erikson (1959) views it as a time of identity-crisis, in which the child struggles for a stable sense of self. Psychoanalytic theory postulates that the burgeoning sexual desires of puberty spark a resurgence of oedipal conflicts for the boy and pre-oedipal pressures for the girl (Blos 1962; A. Freud 1958). To establish mature cross-sexual relationships in adulthood, the child must resolve these conflicts during adolescence. In the interim, the physiological changes of puberty and the increase in sexual desire challenge the child's view of himself in fundamental ways. Both his body-image and his self-image radically change.

Sociologists (Davis 1944) traditionally characterize adolescence as a period of physical maturity and social immaturity. Because of the complexity of the present social system, the child reaches

199

physical adulthood before he is capable of functioning well in adult social roles. Adolescence becomes extremely difficult because the new physical capabilities and new social pressures to become independent coincide with many impediments to actual independence, power, and sexual freedom.

The resulting status-ambiguities, that is, the unclear social definitions and expectations, have been seen as engendering a corresponding ambiguity of self-definition. In addition, the need to make major decisions about future adult roles on the basis of what he is like at present further heightens the adolescent's self-awareness and self-uncertainty (Erikson 1959). . . .

Since most work on adolescent disturbance has been clinical in nature, several fundamental questions on the self-image remain to be answered. First, do data support the belief that the adolescent's self-image differs from that of younger children? If so, could one term this difference a "disturbance," that is, a change which would cause the child some discomfort or unhappiness? In this paper we use the word "disturbance" as a milder term than "turmoil," "storm or stress," or "crisis," so that we can encompass less severe changes. It is not meant to imply psychopathology.

Second, if there is an adolescent self-image disturbance, when does it begin? This question is crucial to the evaluation of certain theoretical notions. Erikson (1959) tells us that the adolescent must deal with the issues of a career decision and the establishment of his own family. While these concerns may be salient to the eighteen- or nineteen-year-old, they do not concern the twelve-year-old. Conversely, it is the younger adolescent who is confronted with the body-image changes of puberty. This study tries to specify the onset of adolescent self-image disturbance.

Third, if there is an adolescent self-image disturbance, what is the course of its development? Do the problems appearing at the time it is precipitated continue to grow? Do they level off at a higher plane? Or do they decline as the adolescent learns to cope with them?

Finally, if it does exist, what triggers the adolescent disturbance? Typically, the onset of puberty is viewed as the trigger. But perhaps aspects of the social environment are at work.

Self-Image Dimensions

In this paper we adopt Gardner Murphy's (1947) view of the self as "the individual as known to the individual." So conceived, the self-image can be viewed as an attitude toward an object; and, like all attitudes, it has several dimensions (Rosenberg 1965). We shall deal with four of these. In each case, there is reason to think that changes in these dimensions would be disturbing or uncomfortable for the individual.

The first dimension is self-consciousness: it refers to the salience of the self to the individual. As Mead (1934) posited, in an interaction the ordinary individual must take account of others' reactions to himself and his behavior. But people vary in the degree to which the self is an object of attention. Some people are more "task-oriented," i.e., more involved in the situation and less concerned with how they are doing or what others are thinking of them. For others, the self becomes so prominent that the interaction is uncomfortable. Do adolescents show more of this type of uncomfortable self-consciousness than younger children?

The second dimension of the self-image is stability. If an individual must take account of himself as an important part of a situation and if he is unsure of what he is like, then he is deprived of a basis for action and decision. . . . The question is whether this stability is especially shaken during adolescence.

The third dimension is self-esteem, i.e., the individual's global positive or negative attitude toward himself. . . . Is there evidence of self-esteem disturbance during adolescence?

The final dimension deals with the "perceived self." While technically not an integral part of the phenomenal self, there is both theoretical and empirical reason to believe that the perceived self has an extremely important bearing on the self-image, particularly the self-esteem. Mead's (1934) and Cooley's (1912) classic theories emphasized the importance to the individual of his perceptions of how others see him. . . . Our question is whether adolescents are more likely than younger children to see others as viewing them unfavorably?

METHOD

Sample

The data for this analysis were collected from public school children in grades three through twelve in Baltimore City in 1968. A random sample of 2,625 pupils distributed among 25 schools was drawn from the population of third to twelfth grade pupils. Each school in Baltimore City was initially stratified by two variables: (1) proportion of nonwhite students, and (2) median income of its census tract. Twenty-five schools falling into the appropriate intervals were randomly selected. From each school, 105 children were selected by random procedures from the central records.

Each subject was interviewed directly after school in his school. For the elementary school children, objective background information was collected from the parents. Parents were reached either by a five to ten minute telephone interview or, when there was no telephone, by home interview. Almost all parents were extremely cooperative and in only 60 cases were we unable to locate the parent or conduct the interview.

Measures

Indexes were developed to measure the four aspects of the self-image discussed above. . . . "Self-consciousness" is based on a seven-item Guttman Scale. (Example: "If a teacher asked you to get up in front of the class and talk a little bit about your summer, would you be very nervous, a little nervous, or not at all nervous?") "Stability of self" is indexed by a five-item Guttman Scale. (Example: "A kid told me: 'Some days I like the way I am. Some days I do not like the way I am.' Do your feelings *change* like this?")

Since the self-esteem dimension is central, this concept was measured in two ways. First, we ascertained the individual's general, overarching feeling toward himself through a series of general questions; we call this the global measure of self-esteem. For this purpose, a six-item Guttman Scale was used. (Example: "Everybody has some things about him which are good and some things about him which are bad. Are most of the things about you good, bad, or are both about the same?"). . . .

In this study, the specific approach to self-esteem measurement is based on the individual's average self-assessment on the following eight characteristics: being smart, good-looking, truthful or honest, good at sports, well-behaved, hard-working in school, helpful, and good at making jokes.

The individual has many perceived selves since he interacts with many types of people who evaluate him. Some of these perceived selves were investigated by our asking these children what they believed the following people thought of them: their parents, their teachers, children of the same sex, and children of the opposite sex.

RESULTS

The Disturbance

Does adolescence produce a disturbance in the child's self-picture? Table 16.1 clearly suggests that the emergence of self-image problems in adolescence is no myth and that these problems occur early in adolescence. In general, self-image disturbance appears much greater in the twelve- to fourteen-year-old age group than in the eight- to eleven-year-old group.

In contrast to younger children, the early adolescents (twelve- to fourteen-year-olds) show a higher level of self-consciousness, greater instability of self-image, slightly lower global self-esteem, lower specific self-esteem, and a more negative "perceived self" (that is, they are less likely to think that parents, teachers, and peers of the same sex view them favorably). The assumption that such changes are likely to be disturbing is consistent with the fact that early adolescents also show a higher level of depressive affect than do the younger children (table 16.1). The only area showing improvement in early adolescence involves the opposite sex: children see themselves as better liked by the opposite sex as they grow older.

While the early adolescents are more self-conscious and have a more unstable self-image, this self-consciousness appears to decline somewhat in later adolescence and the self-image becomes somewhat more stable. However, even in late adolescence, the subjects manifest greater self-consciousness and instability than do the eight- to eleven-year-old children.

Only for global self-esteem is there an improvement in later adolescence marked enough for the youngsters from age fifteen up to score more favorably than the eight- to eleven-year-olds. The older adolescents show higher global self-esteem than both the young children and the early adolescents. . . .

Although global self-esteem feelings decline only slightly in early adolescence and rise conspicuously in later adolescence, this pattern is not true of self-esteem based on those specific qualities we

Table 16.1. Children's Self-Ratings by Age

	Age			
	Median Scores			Total:
	8–11 (N = 819)[a]	12–14 (N = 649)	15+ (N = 516)	Median χ^2 Test[b]
Self-consciousness (Low score = high self-consciousness)	3.8***	3.0	3.2	p < .001
Stability of the self-image (Low score = high instability)	2.6***	2.1	2.3	p < .001
Self-esteem (global) (Low score = low self-esteem)	4.0	3.8***	4.4	p < .001
Self-esteem (specific) (High score = unfavorable rating)	3.6***	5.0	5.0	p < .001
Perceived self (High score = unfavorable rating)				
Perceived opinion of parents	4.8**	5.1	5.1	p < .01
Perceived opinions of teachers	3.2***	3.4	3.4	p < .001
Perceived opinions of peers of the same sex	1.6***	1.8**	2.0	p < .001
Perceived opinions of peers of opposite sex	2.4***	2.1**	2.0	p < .001
Depressive affect (High score = high depression)	2.3***	2.9	3.0	p < .001

* = $p < .05$ for adjacent age groups according to median χ^2 test;
** = $p < .01$;
*** = $p < .001$.
Tests between adjacent age groups are not entirely appropriate, in part, because of the nonindependence of comparisons (i.e., the 12–14 age category is compared with each of the other age groups). Since the test affords some indication of how seriously to take the observed differences, however, it is included for convenience.
    [a]For missing data, total cases are reduced accordingly.
    [b]Siegel (1956:111–116, 179–184).

have considered, such as intelligence, honesty, diligence, and good behavior. If one simply averages the self-ratings on these qualities, he will find a relatively sharp decline between childhood and early adolescence; and this lowered self-evaluation continues into later adolescence.

While one may reasonably assume that the lowered self-evaluations of these specific qualities indicate some degree of self-image disturbance, this conclusion is not certain. For, as William James (1890) long ago noted, it is not simply a question of how favorably the individual judges himself, but also how much he has staked himself on a particular quality. For example, an adolescent may agree that he is poor at sports or is plain; but if he cares little about these qualities, he will not be disturbed by their lack. Thus, only a low self-rating on a quality that is valued highly is likely to be experienced as disturbing.

To take account of self-values, we asked our respondents how much they cared about each of these qualities, i.e., how important they were to them. Table 16.2 deals solely with children who care "very much" about whether they are smart, good-looking, helpful, etc. It is among these children that one would expect an unfavorably self-rating on a quality to be psychologically upsetting (Rosenberg 1965: ch. 13).

With respect to this criterion, table 16.2 indicates that the early adolescents (twelve to fourteen) have a consistently lower self-image

Table 16.2 Proportion Rating Selves Very Favorably on Each Characteristic among Those Who Care "Very Much" about That Characteristic, by Age

Respondent Rates Self Very Favorably on Following Qualities	Age			Total: χ^2 Test
	8–11	*12–14*	*15 or Older*	
Smart	26%***	9%	5%	$p < .001$
	(547)	(366)	(244)	
Good-looking	20%*	13%*	6%	$p < .001$
	(258)	(197)	(121)	
Truthful or honest	54%***	38%	38%	$p < .001$
	(527)	(424)	(320)	
Good at sports	50%	46%	42%	$p < .001$
	(339)	(266)	(163)	
Well-behaved	46%***	31%	39%	$p < .001$
	(474)	(332)	(239)	
Work hard in school	71%***	50%**	39%	$p < .001$
	(494)	(373)	(231)	
Helpful	60%***	46%	46%	$p < .001$
	(506)	(329)	(225)	
Good at making jokes	49%	40%	46%	$p < .05$
	(151)	(53)	(28)	

* $= p < .05$ for adjacent age groups according to χ^2 test;
** $= p < .01$;
*** $= p < .001$.

Tests between adjacent age groups are not entirely appropriate, partly because of non-independence of comparisons. Since the test affords some indication of how seriously to take the observed difference, however, it is included for convenience.

than the younger children (eight to eleven); i.e., they are less likely to rate themselves very favorably on the qualities they consider important. In some cases, such as being "good at making jokes," the differences are minor; for others they are large. On the other hand, there is little consistent difference between early and later adolescents in this regard. The consistent and clear age difference appears between childhood and early adolescence, with the early adolescents less likely to say they are performing well with respect to their self-values.

It may be contended that the lower self-ratings on these qualities simply reflect the fact that adolescents are more "realistic" while the younger children tend to "inflate" their self-qualities. . . . Perhaps the adolescent does become more realistic about what he is like, but this does not mean that the adjustment to reality is not distressing for him. . . .

To summarize, the results show a general pattern of self-image disturbance in early adolescence. The data suggest that, compared to younger children, the early adolescent has become distinctly more self-conscious; his picture of himself has become more shaky and unstable; his global self-esteem has declined slightly; his attitude toward several specific characteristics which he values highly has become less positive; and he has increasingly come to believe that parents, teachers, and peers of the same sex view him less favorably. In view of these changes, it is not surprising that our data show early adolescents to be significantly more likely to be psychologically depressed.

The course of self-image development after twelve to fourteen is also interesting. In general, the differences between early and late adolescence are not large. There is improvement in self-consciousness, stability, and especially global self-esteem, but no improvement in assessment of specific qualities or in the perceived self. The main change occurs almost always between the eight- to eleven-year-old children and the twelve- to fourteen-year-old children. . . .

SUMMARY AND DISCUSSION

This cross-sectional study has investigated several dimensions of self-image development in 1,917 urban school children in grades three through twelve. A definite disturbance of the self-image has been shown to occur in adolescence, particularly early adolescence. In some respects this disturbance appears to decline in later adolescence, while along other dimensions it persists. . . .

During early adolescence, compared to the years eight to eleven, the children exhibited heightened self-consciousness, greater instability of self-image, slightly lower global self-esteem, lower opinions of themselves with regard to the qualities they valued, and a reduced conviction that their parents, teachers and peers of the same sex held favorable opinions of them. They were also more likely to show a high depressive affect. . . .

Knowledge about self-concept development is still pretty much an unknown land in social psychology. Our sample tells us something about what differences appear between the ages of eight and eighteen, but there is little information about development before and after these years. Whether the level or type of self-image disturbance which develops in early adolescene persists in adult life or changes in a positive or negative direction is still unknown. Nor does our study reveal the more dynamic processes of self-image change. That would require a long-term panel study. Given the importance of the self-concept to the individual, we hope that the required research will be forthcoming.

REFERENCES

Blos, P. 1962. *On Adolescence: A Psychoanalytic Interpretation*. New York: Free Press.

Cooley, Charles H. 1912. *Human Nature and the Social Order*. New York: Scribner.

Davis, Kingsley. 1944. Adolescence and the social structure. *Annals of the American Academy of Political and Social Science* 236:8–16.

Erikson, E. H. 1959. Identity and the life cycle. *Psychological Issues* 1:11–171.

Freud, Anna. 1958. Adolescence. *Psychoanalytic Study of the Child* 13:255–278.

Hall, G. S. 1904. *Adolescence: Its Psychology and its Relations to Physiology, Anthropology, Sociology, Sex, Crime, Religion and Education*. New York: Appleton.

James, W. 1950. *The Principles of Psychology*. New York: Dover (copyright, 1890 by Henry Holt and Company).

Mead, George Herbert. 1934. *Mind, Self and Society*. Chicago: University of Chicago Press.

Murphy, Gardner. 1947. *Personality*. New York: Harper.
Rosenberg, Morris. 1965. *Society and the Adolescent Self-Image*. Princeton, N.J.: Princeton University Press.
Siegel, Sidney. 1956. *Nonparametric Statistics for the Behavioral Sciences.* New York: McGraw-Hill.

Adolescence
Relations with Others

Seventeen

Changes in Family Relations at Puberty

Laurence D. Steinberg

Adolescence is a time of important shifts in family relationships. Interaction patterns, communication styles, and decision-making processes all gradually shift as the maturing adolescent demands, and is generally given, more autonomy in the home. In this selection, Laurence Steinberg demonstrates that *puberty*—the physical changes of early adolescence—may serve as a signal to the adolescent and to his parents that the family is ready for a change. Interestingly, Steinberg finds, as did Hetherington, Cox, and Cox in their study of divorce (see selection 12), that tension may be greatest in the mother-son relationship.

THE PURPOSE of the present investigation was to describe and establish possible influences on transformations in parent-child interaction at the time of the physical changes of puberty. The notion that relationships in the family change sometimes during early adolescence has received a great deal of theoretical and popular attention, but comparatively little empirical investigation. Proponents of psychodynamic views of adolescence, for example, have suggested that modifications in family relations at adolescence result in part from changes in the appearance and basic competencies of the child at puberty. Other writers (cf. Erikson 1956; Simmons, Rosenberg, and Rosenberg 1973) place more emphasis on the mediating role of the adolescent's identity and how changes in the child's appearance may influence relations with significant others. Nevertheless, how these fundamental changes in the young adolescent affect relationships in the family system remains empirically unexplored.

In an earlier paper (Steinberg and Hill 1977), the present author reported preliminary data suggesting that, consistent with Jacob's

211

(1974) findings, early adolescence may well be a time of changes in family relations and, furthermore, that patterns of family relations are related to specific changes in the physical appearance of the adolescent. The present investigation reports data from a one-year longitudinal follow-up of the sample of families whose interaction patterns were described in that earlier work.

Steinberg and Hill had reported that the behaviors of male adolescents and their parents in a structured interaction situation are related to the adolescent's physical maturity, independent of his age or formal reasoning abilities. Sons who are more physically mature interrupt their parents more often and explain themselves less. Additionally, adolescents who are near the apex of their physical growth spurts yield to their mothers less often than do prepubertal or late pubertal boys.

Parental behaviors, in contrast, are more often curvilinearly related to the adolescent's physical maturity. While parental explanations are less frequent in families with apex pubertal sons than in families with prepubertal or late pubertal boys, parental interruptions of sons are higher during the apex pubertal period than before or after. Patterns of interaction are more rigidly structured in families with sons near the pubertal apex than in families with less or more mature adolescents.

Using the structured interaction methodology described by Ferreira (1963), and used by Steinberg and Hill and by Jacob, indexes of interpersonal conflict and assertiveness (attempted interruptions), dominance (successful interruptions; talking time), explanations, interaction structure, and actual influence over family decisions were constructed. Changes in these variables were examined in relation to the physical maturation of early adolescent boys. These relations were explored in order to delineate more fully whether and how family relationships are modified around the time of the adolescent boy's pubertal onset.

METHOD

Sample

One-hundred-thirty early adolescent boys and their parents were contacted through local schools, scouting, and recreational programs. Selection criteria were: (1) that the participating adolescent be the oldest male child in a two-parent family, and (2) that one of the parents hold a white-collar position in business, management, or the professions. Of the 130 families contacted, 82 (63 percent) were excluded on the basis of one of the screening criteria; of the remaining families, 17 (35 percent) were unwilling to participate, and 31 (65 percent) agreed to take part in the research program. All but one of the family groups consisted of an adolescent and his natural parents; the one exception was a triad composed of an adolescent, his mother, and his stepfather. All but one of the families were white. The boys ranged in age from 11 years, 3 months to 14 years, 0 months (X = 12 years, 7 months) at the time of the first visit.

Procedure

Families were visited in their homes three times by two observers; in most cases, one of the observers was male and one female. Six months elapsed between visits; with the exception of two cases, no families were seen by the same observer twice. Two families completed the first visit only; two additional families completed only the first and second visit. The procedure for each of the visits was identical. Each father-mother-son triad engaged in a family decision-making task modeled after the "unrevealed differences" procedure (Ferreria 1963). Each individual was instructed to note his own first and second preferences on a set of multiple-choice questions on topics of interest to most middle-class families. The questions concerned such issues as selecting a restaurant, planning a vacation, and deciding how to spend an afternoon together. Three alternative forms of decision-making questionnaires were employed; the forms were assigned to families on a random basis in a random order, and no family received the same form twice. After the family's choices had been recorded on individual questionnaires, one of the observers, using standard instructions, asked the group to discuss the set of questions and come to a "family decision" on first and second preferences for each one. This discussion was taped and took place in the absence of any experimenter.

During the home visit, the observers rated the adolescent along three dimensions (face, body proportion, coordination), and used these ratings to make a global assessment of the adolescent's physical status. The boys were classified as either *prepubertal, early pubertal, apex pubertal, postapex,* or *late pubertal.* The five-point scale used for this classification was constructed on the basis of findings reported by Tanner (1962) and Stolz and Stolz (1951) and in consultation with local pediatricians. A boy was classified as *prepubertal* if he exhibited no signs of change in appearance associated with puberty; the prepubertal boy's face is round in shape, his body is well proportioned but lacking in muscular definition, and his gait is co-ordinated but childlike. The *apex pubertal* adolescent is near the apex of his growth sput; his face is angular, but lacking in hair or acne, his limbs are not in proportion with his trunk, and his gait is gawky and uncoordinated. The *late pubertal* adolescent is well past his growth apex; he has a small amount of facial hair, and in some cases acne, his body is well proportioned and his muscles are well defined, his gait is coordinated and adultlike. The *early pubertal* and *postapex* classifications were reserved for boys whose physical maturity status lies between the three anchor points. The physical status classifications were made before the observers heard any family interaction and without the knowledge of any family member.

Coding

Dependent variables. The taped interactions were coded independently and reliably by trained judges who were blind to the age, intellectual level, and physical maturity of the participating son. The tapes were coded for the following variables: tape length (in seconds), talking times (in seconds), attempted interruptions for each dyad (an attempted interruption was defined as an attempt by one individual to make an assertion or to ask a question while another individual was still speaking), yielding to attempted interruptions for each dyad (yielding was noted when an individual abruptly stopped talking before his idea was completed (cf., Mishler and Waxler 1968), explanations (an explanation was defined as an attempt to give a reason or reasons for an assertion or opinion), and the sequence of speakers.

In order to make meaningful between-family comparisons, the data were transformed in the following ways: individual talking times were converted to proportions of total talking time; interruptions were divided by the talking times of the interrupting and interrupted individuals; successful interruptions were divided by total interruptions to form success ratios; and the sequence of speakers was transformed into proportions of dyadic interchange. These proportions were converted into "structure" scores indicating the degree to which discussions were dominated by interchange of a certain kind; higher scores indicate greater structure (Haley 1964).

Each family member's actual influence over the final decision was determined in two different manners. An *influence* score was calculated based on the number of times an individual's first preference corresponded to the family's first or second preferences. Preliminary analyses indicated that few interaction variables were systematically affected by either the repeated measures design or by differences between the questionnaire forms.

RESULTS

Physical Maturation over the Course of One Year. During the one-year period of investigation, 21 boys (72 percent) matured physically and 8 (28 percent) did not. (One boy in each of these groups was observed only twice over a six-month period.) Twenty-seven boys, therefore, had two periods of potential physical maturation (from Time 1 to Time 2 and from Time 2 to Time 3) and two boys had only one such period (Time 1 to Time 2). Each adolescent's six-month periods were classified on the basis of the presence and/or type of physical maturation into one of eight categories, which appear in table 17.1 along with the distribution of adolescents across these groups.

Table 17.1. Distribution of Adolescents Across Physical Change Groups for First and Second Six-Month Periods

	Group	First Period (Time 1–Time 2)	Second Period (Time 2–Time 3)
(1)	Prepubertal—Prepubertal	2	0
(2)	Prepubertal—Early Pubertal, Apex Pubertal	5	2
(3)	Early Pubertal—Early Pubertal	4	3
(4)	Early Pubertal—Apex Pubertal, Postpex	3	3
(5)	Apex Pubertal—Apex Pubertal	5	2
(6)	Apex Pubertal—Postapex, Late Pubertal	1	5
(7)	Postapex—Postapex	3	3
(8)	Postapex—Late Pubertal	4	2
(9)	Late Pubertal—Late Pubertal	2	7
	Total	29	27

Effects of Physical Maturation on Family Relations. Difference scores on all dependent variables were calculated for each six-month period for every family. Families with adolescents who matured during a six-month period were then matched with families whose adolescents began the six-month period at the *same* physical status but *did not mature.* Thus, (see table 17.1) Group 1 (Prepubertal—Pubertal) was matched with Group 2 (Prepubertal—Pubertal), Group 3 (Early Pubertal—Early Pubertal) with Group 4 (Early Pubertal—Pubertal), Group 5 (Apex Pubertal—Apex Pubertal) with Group 6 (Apex Pubertal—Postapex) and Group 7 (Postapex—Postapex) with Group 8 (Postapex—Late Pubertal). Within each contrast, the Mann-Whitney Test was used to examine differences between the magnitudes of the change scores.

Summary of Changes in Family Interaction over the Pubertal Cycle

Interruptions. Over the pubertal cycle, adolescents come to interrupt their parents, particularly their mothers, more frequently. While all parties interrupt each other at about equal rates during the prepubertal period, the adolescent begins to interrupt his mother more during early puberty ($U = 7$, $p < .08$). During this same time there is an increase in the degree to which mothers interrupt their sons as well, although it is not statistically significant. This trend reverses, however, during the apex pubertal period ($U = 0$, $p < .005$), but the level with which the mother interrupts the son never declines to the prepubertal level.

Fathers also show an overall increase in the rate with which they interrupt their sons ($U = 2$, $p < .02$), but this increase does not take place *until* the apex pubertal period, the time during which the mothers' interruptions of their sons decline. Patterns of interruptions over the pubertal cycle are shown in figure 17.1.

Yielding. Comparatively few significant changes take place in yielding during the pubertal cycle. During the prepubertal peiod, the adolescent yields to his parents far more than they yield to him. While the rate with which he yields to his father does not appear to change during the early pubertal period, it does decline, although not significantly, during the postapex stage. There is also a nonsignificant decline in the rate with which the adolescent yields to his mother, but

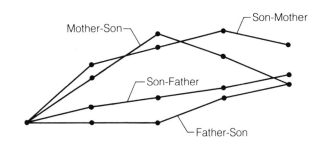

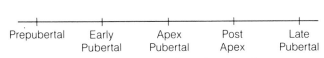

Figure 17.1. Interruptions over the pubertal cycle.

this does occur during the apex pubertal period. This rate then increases into late puberty.

No changes appear to take place in the level with which the father yields to the adolescent, suggesting that this may be a particularly stable dominance relationship. Mothers do, however, show a decline in the degree to which they yield to their sons during the pubertal apex period ($U = 0$, $p < .05$) followed by an increase in this rate during the postapex period ($U = 0$, $p < .10$). These results suggest that the mother-son dominance relationship may be in a state of transition at the time of the pubertal apex. Patterns of yielding over the pubertal cycle are shown in figure 17.2.

Talking Time. Virtually no changes in relative talking times appear to take place during the pubertal cycle. Fathers talk more than mothers or sons, who talk at about equal rates. This suggests that this index of dominance is especially stable in middle-class families.

Influence. Changes in absolute and relative levels of influence appear to occur during the pubertal cycle. Sons gain influence relative to their parents during the earlier part of the cycle, although this increase is not significant. The adolescent's level of influence is relatively stable during the later part of the cycle.

Mothers appear to lose influence during the apex period ($U = 7$, $p < .09$), and this decline continues into the late pubertal

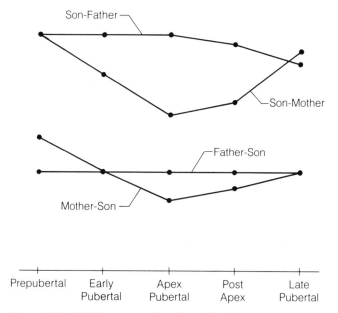

Figure 17.2. Yielding over the pubertal cycle.

period ($U = 5$, $< .06$). Fathers' levels of influence, by contrast, appear to change very little, either relatively or absolutely. Thus, the apex pubertal period—the time during which the mother-son relationship appears to change—is also the time during which the mother's influence begins to decline. Patterns of influence over the pubertal cycle are shown in figure 17.3.

Explanations. Decreases in explanations on the part of all family members occur over the pubertal cycle, although this trend for adolescents is not significant. It should be noted, however, that any decline in adolescent explanations may be due in part to the effects of repeated participation in the task. The overall pattern of change in explanations does suggest, however, an equalization in rates of explaining over the pubertal cycle. Patterns of explanations over the pubertal cycle are shown in figure 17.4.

Interaction Structure. The apex pubertal period appears to be a time of greater structure in interaction patterns than previous ($U = 7$, $p < .09$) or subsequent ($U = 4$, $p < .06$) periods, suggesting that families may introduce greater levels of structure into their interaction

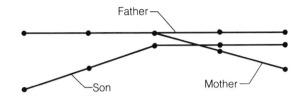

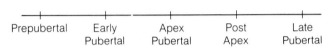

Figure 17.3. **Influence over the pubertal cycle.**

during times of changing relationships. Interaction structure over the pubertal cycle is shown in figure 17.5.

DISCUSSION

Both Steinberg and Hill (1977) and Jacob (1974) have reported data indicating that early adolescence appears to be a time of transfor-

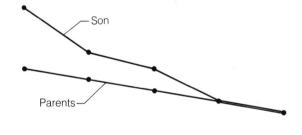

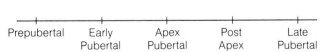

Figure 17.4. **Explanations over the pubertal cycle.**

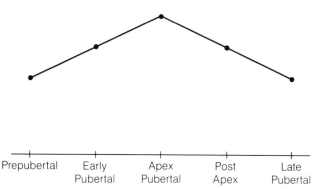

| Prepubertal | Early Pubertal | Apex Pubertal | Post Apex | Late Pubertal |

Figure 17.5. Interaction structure over the pubertal cycle.

mation in familial relationships and, in particular, the adolescent son's position in the family vis-à-vis his mother. The findings presented here both substantiate and further elaborate upon this notion.

Before the adolescent boy's pubertal onset, the influence hierarchy is dominated by the parents, who share the top position equally; by the time the son has entered the late pubertal stage, he has risen within the hierarchy, but at the expense of his mother: she now ranks third in influence, behind her son. Similar father-son-mother influence hierarchies have been found by other researchers studying parent-child triads in families with older adolescent boys who, by virtue of their age, are likely to be physically mature (Strodtbeck 1958; Jacob 1974).

Apparently, the influence loss by the mother is preceded by a struggle between her and her son. As the adolescent begins to mature physically, he attempts to interrupt his mother more frequently during family discussion. These increased attempts are mirrored by increased attempts on the mother's part to interrupt her son. Both behaviors increase steadily in frequency from the time of the adolescent's pubertal onset until his pubertal apex. The adolescent's attempts at assertion are not successful, however; his mother refuses to yield to his interruptions. In fact, as his attempts at assertion increase in frequency, so do her refusals to defer to him. This conflict is reflected in an increase in the degree to which the interaction follows a more structured pattern, which also rises from the time of the pubertal onset until the pubertal apex.

It is not until the adolescent has reached his pubertal apex and until the mother-son conflict has become full blown that the adolescent's father intervenes. The period immediately following the pubertal apex is characterized by an increase in his interruptions of his son. The father is apparently more effective in this endeavor than was his wife; his son soon becomes far less assertive toward both parents.

The struggle between mother and son comes to an end soon after the pubertal apex. Accompanying his decreased assertiveness toward her is a comparable decrease in her interruptions of him. This decreased conflict is reflected in a decrease in interaction structure as well. Consistent with this, as the adolescent enters the late pubertal period, is the mother's increased deference toward her son; this may be a sign that she has conceded to him somewhat; mothers continue to lose influence through the latter portion of the adolescent's pubertal cycle.

The changes in familial relations described above appear to be directly linked to the physical maturation of the adolescent male. Whether this linkage is rooted in revivified oedipal conflict or in changes in the adolescent as a social stimulus—to himself and to his parents—cannot be determined on the basis of these data. What is worthy of speculation, however, is why it is the adolescent's mother who is the target of his ascendance.

As examination of the cross-lag correlations between *son interrupts mother* and *mother interrupts son* using the subsample of boys who matured from the pubertal onset to the apex pubertal period ($N = 6$), the time during which the struggle apparently begins, was undertaken to further explore the development of the adolescent–mother conflict. While the correlation between *son interrupts mother* during early puberty and *mother interrupts son* during the apex period is statistically significant ($r = .86$, $p < .05$), the correlation between *mother interrupts son* during early puberty and *son interrupts mother* during the apex period is close to zero ($r = -.07$, N.S.).

This suggests that it is the adolescent who initiates the conflict and not his mother. The male adolescent, as he matures, may learn a set of assertive behaviors in interactions with his friends; Savin-Williams (1977), for example, reports that dominance in male adolescent peer groups is associated with physical maturity rather than chronological age or mere size. This assertiveness may be incorporated into the adolescent's developing sense of identity and experimented with at home. It may be the case that in middle-class American families, mothers are easier targets for this experimentation than are fathers.

An intricate set of relational transformations in the family is thus set in motion, either directly or indirectly, by the physical changes of puberty. The adolescent-parent system appears to move through a period of disequilibrum and adaptation as the male child changes physically. In the final analysis, however, it may be the adolescent's mother, and not the child himself, whose position in the family changes the greatest as a result of the adolescent's maturation.

Whether this temporary period of apparent conflict between the adolescent and his mother is evidence supportive of the psychoanalytic view of adolescence as an inevitable period of "storm and stress" is open to question, however. The "storm and stress" may be unconscious, given the generally positive picture of family relations reported by adolescents and their parents (Douvan and Adelson 1966; Offer 1969; Kandel and Lesser 1972). The fact that it may be unconscious, however, should not obscure its possible significance. The conflict may very well be a necessary element in the socialization of interpersonal assertiveness in the male adolescent.

REFERENCES

Douvan, E. and J. Adelson. 1966. *The Adolescent Experience*. New York: Wiley.

Erikson, E. 1965. The problem of ego identity. *J. American Psychoanalytic Association* 4:56–121.

Ferreira, A. J. 1963. Decision-making in normal and pathological families. *Archives of General Psychiatry* 8:68–73.

Haley, J. 1964. Research on family patterns: an instrument measurement. *Family Process* 3:41–65.

Jacob, T. 1974. Patterns of family conflict and dominance as a function of age and social class. *Developmental Psychology* 10:1–12.

Kandel, D. and G. Lesser, 1972. *Youth in Two Worlds*. San Francisco: Jossey-Bass.

Mishler, E. and N. Waxler. 1968. *Interaction in Families*. New York: Wiley.

Offer, D. 1969. *The Psychological World of the Teenager*. New York: Basic Books.

Savin-Williams, R. 1977. Personal communication. University of Chicago.

Simmons, R., F. Rosenberg, and M. Rosenberg. 1973. Disturbance in the self-image at adolescence. *American Sociological Review* 38:553–568.

Steinberg, L. D. and J. P. Hill. 1977. Family interaction in early adolescence. Paper presented at the biennial meeting of the Society for Research in Child Development, New Orleans.

Stolz, H. and L. Stolz. 1951. *Somatic Development of Adolescent Boys.* New York: Macmillan.

Strodtbeck, F. 1958. Family interaction, values, and achievement. In D. McClelland, A. Baldwin, U. Bronfenbrenner, and F. Strodtbeck, eds., *Talent and Society,* Princeton: Van Nostrand.

Tanner, J. 1962. *Growth at Adolescence,* 2nd ed. Springfield, Ill.: Charles C. Thomas.

Eighteen

Breakups before Marriage: The End of 103 Affairs

Charles T. Hill, Zick Rubin, and Letitia Anne Peplau

Although a great deal is known about dating and interpersonal attraction during the late adolescent years, few researchers have looked at the process of "breaking up." In this selection, Hill, Rubin, and Peplau report on a major study of the breakups of college student couples. Among the many interesting findings of this study is that women appear to be somewhat more in control of these dating relationships than men: women are more likely to initiate breakups, find it easier to remain "friends" after the relationship ends, and find it easier to cope following the breakup. This suggests that some of our commonsense notions about which sex is the "romantic" one may not be entirely accurate.

FOR ALL THE CONCERN with the high incidence of divorce in contemporary America, marital separation accounts for only a small proportion of the breakups of intimate male-female relationships among American couples. For every recorded instance of the ending of a marriage, there are many instances, typically unrecorded, of the ending of a relationship among partners who were dating or "going together." Such breakups before marriage are of fundamental importance to an understanding of marital separation for two major reasons.

First and foremost, breakups before marriage play a central role in the larger system of mate selection. In an ideal mate selection system, all breakups of intimate male-female relationships might take place before marriage. Boyfriends and girl friends who are not well suited for each other would discover this in the course of dating and would eventually break up. In practice, however, the system does not

224

achieve this ideal. Many couples who subsequently prove to be poorly suited for marrying each other do not discover this until after they are married. In many other instances, couples may be aware of serious strains in their relationship but nevertheless find themselves unable or unwilling to break up before marriage. Many future sources of marital strain may be totally unpredictable at the time that couple decides to get married; individuals' needs and values may change over the course of time in ways that could not have been anticipated initially. Nevertheless, it is possible that the selection system could be made to operate more efficiently than it currently does. Although the psychic cost of a premarital breakup is often substantial, by breaking up before marriage couples might spare themselves the much greater costs of breaking up afterward.

Second, breakup before marriage may provide a revealing comparison against which to view marital breakup. Many of the psychological bonds of unmarried couples resemble those of married couples. Thus the requirements and difficulties of "uncoupling" in the two cases may show similarities (see Davis 1973). On the other hand, breakup before marriage takes place in a very different social context from that of divorce. The ending of a dating relationship is relatively unaffected by factors that play central roles in divorces—for example, changes in residence, economic arrangements, child custody, legal battles, and stigmatization by kin and community. Thus the examination of breakups before marriage may be helpful in untangling the complex of psychological and social factors that influence divorce and its aftermath. . . .

In this paper, we report on breakups before marriage among a large sample of dating couples in the 1970s. Our data are primarily descriptive: How were those couples who broke up over a two-year period different from those who stayed together? What were the reasons for the breakups, as perceived by the former partners themselves? What were the central features of the breaking-up process: its precipitating factors, its timing, and its aftermath? We pay special attention to the two-sidedness of breaking up: the frequent differences in the two partners' perceptions of what is taking place and why, the pervasive role differentiation of breaker-upper and broken-up-with, and the possibility that there are important differences between men's and women's characteristic orientations toward breaking up before marriage.

THE RESEARCH CONTEXT

In the spring of 1972, for a longitudinal study of dating relationships (Rubin, Peplau, and Hill, 1976), we sent a letter to a random sample of 5,000 sophomores and juniors, 2,500 men and 2,500 women, at four colleges in the Boston area. The colleges, chosen with a view toward diversity, included a large private university (2,000 letters) and a small private college, a Catholic university, and a state college for commuter students (1,000 letters per school). Each student was sent a two-page questionnaire which asked if he or she would be interested in participating in a study of "college students and their opposite-sex relationships." A total of 2,520 students (57 percent of the women and 44 percent of the men) returned this questionnaire. Of these, 62 percent of the women and 54 percent of the men indicated that they were currently "going with" someone. Those who said that they and their partner might be interested in participating in a study were invited to attend a questionnaire session—with their boyfriend or girl friend—either at their own school or at Harvard.

The 202 couples who responded to our invitation, plus an additional 29 couples who were recruited by advertising at one of the four schools, constitute our sample (Hill, Rubin, and Willard, 1972). At the time of the initial questionnaire, almost all participants (95 percent) were—or had been—college students. The modal couple consisted of a sophomore woman dating a junior man. About half of the participants' fathers had graduated from college and about one-fourth of the fathers held graduate degrees. About 44 percent of the respondents were Catholic, 26 percent were Protestant, and 25 percent were Jewish, reflecting the religious composition of colleges in the Boston area. Virtually all of the participants (97 percent) were white; about 25 percent lived at home with their parents, another 35 percent lived in apartments or houses by themselves or with roommates, and 38 percent lived in college dormitories. Almost all of the participants— 97 percent of the women and 96 percent of the men—thought that they would eventually get married, although not necessarily to their current dating partner.

At the beginning of the study, the couples had been dating for a median period of about 8 months—a third for 5 months or less, a third between 5 and 10 months, and third for longer than that. In three-fourths of the couples both persons were dating their partner exclusively, but only 10 percent of the couples were engaged and relatively few had concrete plans for marriage. Four-fifths of the couples had had sexual intercourse, and one-fifth were living together "all or most of the time." Sixty percent were seeing one another every day.

Data Collection

      In addition to the initial questionnaire, a follow-up questionnaire was administered in person or by mail 6 months, 1 year, and 2 years after the initial session. At all points response rates were good. For example, in the one-year follow-up, two-thirds of the initial participants attended questionnaire sessions and another 14 percent returned short questionnaires in the mail. Four-fifths of the original participants returned the 2-year mail questionnaire. To categorize a relationship as intact or broken after two years, we have reports from at least one member of all but 10 of the 231 couples. In all cases, boyfriends and girl friends were asked to fill out the questionnaires individually. They were assured that their responses would be kept in strict confidence and would never be revealed to their partners. They were each paid $1.50 for the initial one-hour questionnaire session and $3.00 for a somewhat longer session one year later. To supplement these data, a smaller number of individuals and couples were interviewed intensively. Of particular relevance to this paper is a series of interviews conducted in the fall of 1972 with 18 people whose relationships ended after they began their participation in the study.

WHICH COUPLES BROKE UP?

By the end of the two-year study period, 103 couples (45 percent of the total sample) had broken up. (Of the remaining couples, 65 were dating, 9 were engaged, 43 were married, 10 had an unknown status, and one partner had died.) The length of time that breakup couples had been dating before ending their relationship ranged from 1 month to 5 years; the median was 16 months. On the basis of data obtained in the initial questionnaire, could these breakups have been predicted in advance?

Measures of Intimacy

Burgess and Wallin (1953) list "slight emotional attachment" as a major factor associated with the ending of premarital relationships. Our data indicate that in general those couples who were less intimate or less attached to one another when the study began were more likely to break up (table 18.1). On the initial questionnaire, compared to couples who stayed together, couples who were subsequently to break up reported that they felt less close and saw less likelihood of marrying each other; they were less likely to be "in love" or dating exclusively, and tended to have been dating for a shorter period of time. The data also indicate, however, that many relationships which were quite "intimate" in 1972 did not survive beyond 1974. For example, over half of the partners in breakup couples felt that they were both in love at the time of the initial questionnaire. Whereas some of the couples who were to break up apparently never

Table 18.1. Factors Contributing to the Ending of a Relationship (Percentage Reporting)

	Women's Reports	Men's Reports	Partner Correlation
Dyadic Factors			
Becoming bored with the relationship	76.7	76.7	.23*
Differences in interests	72.8	61.1	.04
Differences in backgrounds	44.2	46.8	.05
Differences in intelligence	19.5	10.4	.17
Conflicting sexual attitudes	48.1	42.9	.33**
Conflicting marriage ideas	43.4	28.9	.25*
Nondyadic Factors			
Woman's desire to be independent	73.7	50.0	.57**
Man's desire to be independent	46.8	61.1	.55**
Woman's interest in someone else	40.3	31.2	.56**
Man's interest in someone else	18.2	28.6	.60**
Living too far apart	28.2	41.0	.57**
Pressure from woman's parents	18.2	13.0	.33**
Pressure from man's parents	10.4	9.1	.58**

NOTE. Data for those couples for which both man's and woman's reports were available ($N = 77$). Factors labeled "man's" and "woman's" above were labeled as "my" or "partner's" in the questionnaires. Percentages are those citing factor as "a contributing factor" or as "one of the most important factors." Correlations are based on 3-point scales.
*$p < .05$
**$p < .01$

developed much intimacy in the first place, others had a high degree of intimacy that they were unable or unwilling to sustain. . . .

The partners' "love" was a better predictor of the couple's survival than their "liking" for one another, as measured by scales previously developed by Rubin (1970, 1973). This distinction is in accord with the conceptual meaning of the two scales, with love including elements of attachment and intimacy, while liking refers to favorable evaluations that do not necessarily reflect such intimacy. In addition, the women's love for their boyfriends tended to be a better predictor of dating status (point-biserial $r = .32$) than the men's love for their girl friends ($r = .18$). Thus the woman's feelings toward her dating partner may have a more powerful effect on a relationship and/or provide a more sensitive barometer of its viability than do the man's.

Finally, two important measures of couple intimacy were totally unrelated to breaking up: having had sexual intercourse of having lived together. These behaviors apparently reflect a couple's social values at least as much as the depth of their attachment to one another. Having sex or living together may bring a couple closer, but they may also give rise to additional problems such as coordinating sexual desires or agreeing on the division of household tasks.

Relative Degree of Involvement

In addition to "slight emotional attachment," Burgess and Wallin (1953) list "unequal attachment" as a factor underlying breakups. The hypothesis that equal involvement facilitates the development of a relationship was spelled out by Blau:

Commitments must stay abreast for a love relationship to develop into a lasting mutual attachment. . . . Only when two lovers' affection for and commitment to one another expand at roughly the same pace do they tend mutually to reinforce their love. (1964:84)

Our data provide strong support for Blau's hypothesis. Of the couples in which both members reported that they were equally involved in the relationship in 1972, only 23 percent broke up; in contrast, 54 percent of those couples in which at least one member reported that they were unequally involved subsequently broke up. It should be noted, however, that there was a significant association between reporting high intimacy on a variety of measures (e.g., those in table 18.1) and reporting equal involvement.

Similarity and Matching

Probably the best documented finding in the research literature on interpersonal attraction and mate selection is the "birds-of-a-feather principle"—people tend to be most attracted to one another if they are similar or equally matched on a variety of social, physical, and intellectual characteristics and attitudes (Rubin 1973). Evidence for such matching was found among the couples in our study.

Although there is less empirical support for it, some researchers have put forth "sequential filtering" models of mate selection which propose that social and psychological similarities or dissimilarities are recognized and responded to in particular sequences. For example, Kerckhoff and Davis (1962) and Murstein (1971) propose that filtering (i.e., the elimination of mismatches takes place first with respect to social background, physical, and other external or stimulus factors, and later with respect to important attitudes and values (Udry 1971).

The intracouple correlations of the breakup and together groups in our sample reveal that couples were more likely to stay together if they were relatively well matched with respect to age, educational plans, intelligence (measured by self-reported SAT scores), and physical attractiveness (measured by judges' ratings of individual color photographs). On the other hand, there was no suggestion of filtering during the period of study on such other presumably important characteristics as social class (indexed by father's education), religion, sex-role traditionalism, religiosity, or desired family size. It may be surmised that any filtering on such factors had already taken place before the time of our initial questionnaire. . . .

THE PROCESS OF BREAKING UP

Brief synopses of two breakups, taken from among the sample we interviewed, may help to illustrate the process of breaking up. Neither of these cases is presented as typical, but the two illustrate several features that are characteristic of the aggregate findings.

Kathy and Joe had been going together during the school year when she was a sophomore and he was a junior. Both the them agree that Kathy was the one who wanted to break up. She felt they were too tied down to one another, that Joe was too dependent and demanded her exclusive attention—even in groups of friends he would draw her aside. As early as the spring Joe came to feel that Kathy was no longer as much in love as he, but it took him a long time to reconcile himself to the notion that things were ending. They gradually saw each other less and less over the summer months, until finally she began to date someone else. The first time that the two were together after the start of the next school year Kathy was in a bad mood, but wouldn't talk to Joe about it. The following morning Joe told Kathy, "I guess things are over with." Later when they were able to talk further, he found out that she was already dating someone else. Kathy's reaction to the breakup was mainly a feeling of release—both from Joe and from the guilt she felt when she was secretly dating someone else. But Joe had deep regrets about the relationship. For at least some months afterward he regretted that they didn't give the relationship one more chance—he thought they might have been able to make it work. He said that he learned something from the relationship, but hoped he hadn't become jaded by it. "If I fall in love again," he said, "it might be with the reservation that I'm going to keep awake this time. I don't know if you can keep an innocent attitude toward relationships and keep watch at the same time, but I hope so." Meanwhile, however, he had not begun to make any new social contacts, and instead seemed focused on working through the old relationship, and, since Kathy and he sometimes see each other at school, in learning to be comfortable in her presence.

David and Ruth had gone together off-and-on for several years. David was less involved in the relationship than Ruth was, but it is clear that Ruth was the one who precipitated the final breakup. According to Ruth, David was spending more and more time with his own group of friends, and this bothered her. She recalled one night in particular when "they were showing *The Last Picture Show* in one of the dorms, and we went to see it. I was sitting next to him, but it was as if he wasn't really there. He was running around talking to all these people and I was following him around and I felt like his kid sister. So I knew I wasn't going to put up with that much longer." When she talked to him about this and other problems, he said "I'm sorry"—but did not change. Shortly thereafter Ruth wanted to see a movie in Cambridge and asked David if he would go with her. He replied, "No, there's something going on in the dorm!" This was the last straw for Ruth, and she told him she would not go out with him anymore. David started to cry, as if the relationship had really meant something to him—but at that point it was too late. At the time we talked to her, Ruth had not found another boyfriend, but she said she had no regrets about the relationship or about its ending. "It's probably the most worthwhile thing that's ever happened to me in my 21 years, so I don't regret having the experience at all. But after being in the supportive role, *I* want a little support now. That's the main thing I look for." She added that "I don't think I ever felt romantic [about David] —I felt practical. I had the feeling that I'd better make the most of it because it won't last that long."

The Timing of Breakups

If dating relationships were unaffected by their social context, it seems likely that they could end at most any time of the year. But the relationships of the couples in our sample were most likely to break up at key turning points of the school year—in the months of May–June, September, and December–January rather than at other times. This tendency, found for the 103 breakups, is illustrated most dramatically in reports of the ending of all respondents' previous relationships, for which there were more than 400 cases.

This pattern of breakups suggests that factors external to a relationship (leaving for vacations, arriving at school graduation, etc.) may interact with internal factors (such as conflicting values or goals) to cause relationships to end at particular times. For example, changes in living arrangements and schedules at the beginning or end of a semester may make it easier to meet new dating partners (e.g., in a new class) or make it more difficult to maintain previous ties (e.g., when schedules conflict or one moves away). Such changes may raise issues concerning the future of a relationship: Should we get an apartment together? Should we spend our vacation apart? Should I accept a job out of state? Should we get together after vacation? If one has already been considering terminating a relationship, such changes may make it easier to call the relationship off. For example, it is probably easier to say, "While we're apart we ought to date others" than it is to say, "I've grown tired of you and would rather not date you anymore." If one is able to attribute the impending breakup to external circumstances, one may be able to avoid some of the ambivalence, embarrassment, and guilt that may be associated with calling a relationship off.

The structuring of breakups by the calendar year was also related to another aspect of the breakup process. In the majority of breakups, like the case of Kathy and Joe, the ending was desired more by the partner who was less involved in the relationship (in this instance, Kathy). In a significant minority of cases, however, the breakup was desired more by the more-involved partner (like Ruth), who finally decides that the costs of remaining in the relationship are higher than he or she can bear. We found a strong tendency for the breakups desired by the less-involved partner to take place near the end or beginning of the school year or during the intervening summer months—71.1 percent April–September vs. 28.9 percent October–March. The breakups desired by the more-involved partner, in contrast, were relatively more likely to take place during the school year—59.1 percent October–March vs. 40.9 percent April–September ($X^2 = 5.68$, $p < .02$). The summer months are, of course, times when college student couples are most likely to be separated because of external factors—for example, returning to homes or jobs in different areas. It seems plausible that less-involved partners would be likely to let their remaining interest in the relationship wane during such periods of separation. Summer separations may also provide a good excuse for the less-involved partner to say good-bye. For the more-involved partner, on the other hand, the period of separation may, if anything, intensify interest in the relationship—"Absence extinguishes small passions and increases great ones" (La Rochefoucauld, quoted in Heider 1958). The more-involved partner may be most likely to end the relationship in response to continuing pain and frustration. As in the case of Ruth, the final break may be precipitated by some "last straw" that occurs while the two partners are still together.

The Two Sides of Breaking Up

The central principle that *there are two sides to every breakup* has both substantive and methodological implications. Very few breakups are truly mutual, with both parties deciding at more or less the same time that they would like to discontinue the relationship. In the present study, 85 percent of the women and 87 percent of the men reported that one person wanted to end the relationship at least somewhat more than the other. Thus in the large majority of cases there are two distinct roles: "breaker-upper" (to be more literary about it, the rejecting lover) and "broken-up-with" (the rejected lover). Identifying these roles is crucial to understanding anything else about a breakup—its underlying reasons, the termination process itself, or its aftermath.

The impact of this role differentiation emerged particularly clearly in self-reports of the emotional aftermath of breaking up (reports available on 1-year follow-up for 31 women, 36 men). Both women and men felt considerably less depressed, less lonely, freer, happier, but more guilty when they were the breaker-uppers than when they were the broken-up-with (for most differences, $p < .01$). For example, whereas Kathy reacted to her breakup with relief, Joe felt deep regret. Indeed, there was a general tendency for the two partners' reactions to a breakup to be inversely related. The freer one partner reported feeling after the breakup, the less free the other partner reported feeling ($r = -.57$, $p < .05$; $N = 15$ cases with both reports). Similar inverse correlations—but of lesser magnitude—characterized the former partners' self-reports of depression, loneliness, and happiness.

A second sense in which there are two sides to every breakup is in the perceptions of the participants; the experience of breaking up is different for each of the two parties involved. For example, although members of couples agreed almost completely on the month in which their relationship finally ended ($r = .98$, $N = 77$), there was only slight agreement on the more subjective question of how gradually or abruptly the ending came about ($r = .24$, $N = 77$). When the former partners were asked to provide their attributions of the causes of the breakup, there was moderate to high agreement on the contribution of nondyadic factors but little or no agreement on factors characterizing the dyad (table 18.1).

One systematic way in which partners' reports disagree concerns who wanted to break up. Although there is a high correlation between men's and women's reports of who wanted the relationship to end ($r = .85$, $N = 76$), there was a systematic self-bias in the reports. There seems to be a general tendency for respondents to say that they themselves, rather than their partners, were the ones who wanted to break up—51.3 percent "I," 35.5 percent "partner," 13.0 percent mutual in the women's reports; 46.1 percent "I," 39.5 percent "partner," 15.0 percent mutual in the men's reports ($N = 76$). Apparently it is easier to accept and cope with a breakup if one views it as a desired outcome (as precipitated by oneself or as mutual) than as an outcome imposed against one's will. A similar self-bias appeared in ratings of factors contributing to the breakup—both men and women cited "my" desire to be independent as more important than "partner's" desire to be independent.

   For some purposes, therefore, it is difficult to speak confidently about *the* breakup, as if it refers to a single, objective set of events. Instead, it seems necessary to attend separately to "his breakup" and to "her breakup," in each instance looking at the matter from the respective partner's point of view—see Bernard's (1972) discussion of "his marriage" and "her marriage." This distinction seems particularly necessary since our data suggest that there may be some systematic differences between men and women in their orientations toward breaking up.

SEX DIFFERENCES IN BREAKING UP

Rubin (1975) has suggested that in respect to dating or premarital relationships in middle-class America today: (a) men tend to fall in love more readily than women, and (b) women tend to fall out of love more readily than men. Evidence for the first proposition has been reviewed elsewhwere (Rubin 1973, 1975). To cite just one datum from the present study, on the initial questionnaire respondents were asked to indicate how important each of a variety of goals was to them as a reason for entering the relationship. Prevailing stereotypes about romantic women to the contrary, men rated the "desire to fall in love" as a significantly more important reason for entering the relationship than did women (4.1 vs. 3.6 on a 9-point scale, $p = .03$). We will review here some of the evidence relating to the second proposition.

Perceived Problems

When participants who had broken up were presented a list of common problems and asked to indicate which had contributed to the breakup (table 18.1), women rated more problems as important than did men ($p < .003$). In particular, more women than men cited "differences in interests," "differences in intelligence," "conflicting ideas about marriage," "my desire to be independent," and "my interest in someone else." Men were only more likely to cite "living too far apart." Although these reports are retrospective and clearly susceptible to distortion, they suggest that women tended to be more sensitive than men to problem areas in their relationship, and that women were more likely than men to compare the relationship to alternatives, whether potential or actual. These tendencies seem consistent with the hypothesis.

Who Precipitates the Breakup?

If women indeed tend to fall out of love more readily than men, we would expect them to be more likely to play the role of breaker-upper and men to play the role of broken-up-with. Combining men's and women's reports of who wanted to break up and classifying a breakup as nonmutual if either partner described it so, we estimate that the woman was more interested in breaking up in 51 percent of the couples, the man in 42 percent, and the breakup was clearly mutual in 7 percent. The participants' reports (200-plus cases for each sex) of breakups in previous opposite-sex relationships (Hill 1974) suggested a similar preponderance of female-initiated breakups. In an earlier study, Rubin (1969) found that 17 of 25 nonmutual breakups among dating couples at the University of Michigan had been initiated by women.

One possible explanation for this datum is that women might have been less involved in these relationships than the men. But that was not the case. Once the relationships had proceeded beyond their early stages, the women were by all indications at least as involved as the men. Combining the two partners' reports before the time of the breakup, women were categorized as the more involved partner in 45 percent of all couples and men as more involved in 36 percent; in 19 percent they were classified as equally involved.

As Blau (1964) has suggested, however, a relationship in which there is unequal involvement will not always be ended by the less involved party: "Whereas rewards experienced in the relationship may lead to its continuation for a while, the weak interest of the less committed or the frustrations of the more committed probably will sooner or later prompt one or the other to terminate it" (p. 84). Our data suggest that Blau's postulated patterns describe a substantial number of breakups precipitated by women. Relationships were often ended by women when they were the less involved partner (67.6 percent of the 34 cases)—like Kathy in the first of the cases presented—and wanted to move on to better alternatives. Relationships were also likely to be ended by women when they were the more involved partner (37.2 percent of the 43 cases)—like Ruth in the second case—and finally abandoned the relationship when they realized and their commitment was not reciprocated. When breakups were precipitated by men, only the first of the two patterns was common. Relationships were frequently ended by men when they were the less involved partner (60.5 percent of the 43 cases), but rarely when they were the more involved (20.6 percent of the 34 cases. These data seem quite consistent with our generalization: Whereas many highly involved women sooner or later find it necessary and possible to terminate the relationship, men seem to find that more difficult.

Staying Friends

If men find it more difficult than women to renounce their love, we might also expect relations between former partners to be more strained after the woman has rejected the man than vice versa. Whereas a rejected woman may be able to redefine her relationship with her former boyfriend from "love" to "friendship"—which, as Davis (1973) notes, is often a euphemism for "acquaintance"—a rejected man may find it more difficult to accomplish such a redefinition. In such cases, staying friends is likely to be impossible. The data support this expectation. A couple was much more likely to stay friends when the man had been the one who precipitated the breakup (70 percent), or when the breakup was mutual (71 percent), than when the woman precipitated it (46 percent) ($X^2 = 5.83$, $p < .06$).

Emotional Aftermath

Our generalization would also suggest that breaking up would be a more traumatic experience for men than for women. Unfortunately, the data available to test this proposition are limited to the 15 couples in which we obtained reports of emotional reactions from both partners on the one-year follow-up. These data suggest that men were hit harder than women by the breakup. Men tended more than women to report that in the wake of the breakup they felt depressed, lonely, less happy, less free, and less guilty. Goethals (1973) presents a clinical discussion of sex differences in reactions to breaking up that seems consistent with these data. In our interviews, we were struck by a particular reaction that appeared among several of the men but not among the women. Some men found it extremely difficult to reconcile themselves to the fact that they were no longer loved and that the relationship was over. Women who are rejected may also react with considerable grief and despair, but they seem less likely to retain the hope that their rejectors "really love them after all."

Two Interpretations

The evidence provides converging support for the notion that women tend to fall out of love more readily than men, just as men may tend to fall in love more readily than women (Rubin 1975). Needless to say, these generalizations are offered as actuarial propositions; they take on importance to the extent that they are informative about aspects of the socialization of the two sexes for close relationships in contemporary America. Two aspects of sex roles may help account for these tendencies.

Simple Economics. Contrary to prevailing stereotypes about romantic and sentimental women, women may be more practical than men about mate selection for simple economic reasons. In most marriages, the wife's status, income, and life chances are far more dependent upon her husband's than vice versa. For this reason, parents in almost all societies have been more concerned with finding appropriate mates for their daughters than for their sons. In "free choice" systems of mate selection like our own, the woman must be especially discriminating. She cannot allow herself to fall in love too quickly, nor can she afford to stay in love too long with the wrong person (Goode 1959). Men, on the other hand, can afford the luxury of being romantic. The fact that a woman's years of marriageability tend to be more limited than a man's also contributes to her greater need to be selective. Waller (1938) put the matter most bluntly when he wrote:

There is this difference between the man and the woman in the pattern of bourgeois family life: a man, when he marries, chooses a companion and perhaps a helpmate, but a woman chooses a companion and at the same time a standard of living. It is necessary for a woman to be mercenary. (p. 243)

Interpersonal Sensitivity. Women are traditionally the social-emotional specialists in most societies, including our own, while men are the traditional task specialists (Parsons and Bales 1955). The emphasis upon social-emotional matters in women's socialization may lead women to be more sensitive than men to the quality of their interpersonal relationships, both in the present and projecting into the future. One possible reflection of women's greater interpersonal sensitivity is the finding, replicated in the present study, that women distinguish more sharply than men between "liking" and "loving" components of interpersonal sentiments (Rubin 1970). Because of greater interpersonal sensitivity and discrimination, it may also be more important for women than for men that the quality of a relationship remain high. Thus women's criteria for falling in love—and for staying in love—may be higher than men's, and they may reevaluate their relationships more carefully.

REFERENCES

Bernard, J. 1972. *The Future of Marriage.* New York: World Book Co.
Blau, P. M. 1964. *Exchange and Power in Social Life.* New York: Wiley.
Burgess, E. and P. Wallin. 1953. *Engagement and Marriage.* Philadelphia: Lippincott.
Davis, M. S. 1973. *Intimate Relations.* New York: Free Press.
Goethals, G. W. 1973. Symbiosis and the life cycle. *British Journal of Medical Psychology* 46:91–96.
Goode, W. J. 1959. The theoretical importance of love. *American Sociological Review* 24:38–47.
Heider, F. 1958. *The Psychology of Interpersonal Relations.* New York: Wiley.
Hill, C. T. 1974. The ending of successive opposite-sex relationships. Ph.D. dissertation, Harvard University.

Hill, C. T., Z. Rubin, and S. Willard. 1972. Who volunteers for research on dating relationships? Manuscript, Harvard University.

Kerckhoff, A. C. and K. E. Davis. 1962. Value consensus and need complementarity in mate selection. *American Sociological Review* 27:295–303.

Murstein, B. I. 1971. A theory of marital choice and its applicability to marriage adjustment. In B. I. Murstein, ed., *Theories of Attraction and Love.* New York: Springer.

Parsons, R. and R. F. Bales. 1955. *Family, Socialization, and Interaction Processes.* Glencoe, Ill.: Free Press.

Rubin, Z. 1969. The social psychology of romantic love. Ph.D. dissertation, University of Michigan.

—— 1970. Measurement of romantic Love. *Journal of Personality and Social Psychology* 16:265–273.

—— 1973. *Liking and Loving: An Invitation to Social Psychology.* New York: Holt, Rinehart, & Winston.

—— 1975. Loving and leaving. Manuscript, Harvard University.

Rubin, Z., L. A. Peplau, and C. T. Hill. 1976. Becoming intimate: the development of male-female relationships. Manuscript.

Undry, J. 1971. *The Social Context of Marriage,* 2d ed. New York: Lippincott.

Waller, W. 1938. *The Family: A Dynamic Interpretation. New York: Dryden.*

Part IV

Adulthood
Mastery and Competencies

Nineteen

Transition to Parenthood
Alice S. Rossi

Along with work (see selection 20), parenthood provides the adult with a role in which mastery and competence can be demonstrated. As Alice Rossi suggests in this selection, however, the transition to this role may be a difficult one for the individual to make. This may be especially true for women who, in most industrialized societies, bear the major responsibility for child rearing. Whether parenthood contributes to, rather than detracts from, the adult's sense of competence may depend to a great extent on the preparation for parenthood the individual receives and on the "fit" between the parenthood role and other roles in the adult's life.

WHAT IS INVOLVED in the transition to parenthood: What must be learned and what readjustments of other role commitments must take place in order to move smoothly through the transition from a childless married state to parenthood? . . .

To get a firmer conceptual handle on the problem, I shall first specify the stages in the development of the parental role and then explore several of the most salient features of the parental role by comparing it with the two other major adult social roles—the marital and work role. . . .

What is unique about this perspective on parenthood is the focus on the adult parent rather than the child. Until quite recent years, concern in the behavioral sciences with the parent-child relationship has been confined almost exclusively to the child. . . .

The very different order of questions which emerge when the parent replaces the child as the primary focus of analytic attention can best be shown with an illustration. Let us take, as our example, the point Benedek makes that the child's need for mothering is *absolute* while the need of an adult woman to mother is *relative* (Benedek 1959). From a concern for the child, this discrepancy in need leads

243

to an analysis of the impact on the child of separation from the mother or inadequacy of mothering. Family systems that provide numerous adults to care for the young child can make up for this discrepancy in need between mother and child, which may be why ethnographic accounts give little evidence of postpartum depression following childbirth in simpler societies. Yet our family system of isolated house-holds, increasingly distant from kinswomen to assist in mothering, requires that new mothers shoulder total responsibility for the infant precisely for that stage of the child's life when his need for mothering is far in excess of the mother's need for the child.

From the perspective of the mother, the question has therefore become: What does maternity deprive her of? Are the intrinsic grat-ifications of maternity sufficient to compensate for shelving or reduc-ing a woman's involvement in nonfamily interests and social roles? The literature on maternal deprivation cannot answer such questions because the concept, even in the careful specification Yarrow (1961) has given it, has never meant anything but the effect on the child of various kinds of insufficient mothering. Yet what has been seen as a failure or inadequacy of individual women may in fact be a failure of the society to provide institutionalized substitutes for the extended kin to assist in the care of infants and young children. It may be that the role requirements of maternity in the American family system ex-tract diversified interests and social expectations concerning adult life. Here, as at several points in the course of this paper, familiar problems take on a new and suggestive research dimension when the focus is on the parent rather than the child. . . .

Parsons' (Parsons and Bales 1955) analysis of the experience of parenthood as a step in maturation and personality growth does not allow for negative outcome. In this view either parents show little or no positive impact upon themselves of their parental-role experi-ences, or they show a new level of maturity. Yet many women, whose interests and values made a congenial combination of wifehood and work role, may find that the addition of maternal responsibilities has the consequence of a fundamental and undesired change in both their relationships to their husbands and their involvements outside the family. Still other women, who might have kept a precarious hold on adequate functioning as adult had they *not* become parents, suffer severe retogression with pregnancy and childbearing, because the reactivation of older unresolved conflicts with their own mothers is not favorably resolved but in fact leads to personality deterioration (Cohen 1966; Rheingold 1964) and the transmission of pathology to their children (Lidz, Fleck, and Cornelison 1965; Rheingold 1964).

Where cultural pressure is very great to assume a particular adult role, as it is for American women to bear and rear children, latent desire and psychological readiness for parenthood may often be at odds with manifest desire and actual ability to perform adequately as parents. Clinicians and therapists are aware, as perhaps many sociologists are not, that failure, hostility, and destructiveness are as much a part of the family system and the relationships among family members as success, love, and solidarity are. . . .

ROLE-CYCLE STAGES

A discussion of the impact of parenthood upon the parent will be assisted by two analytic devices. One is to follow a comparative approach, by asking in what basic structural ways the parental role differs from other primary adult roles. The marital and occupational roles will be used for this comparison. A second device is to specify the phases in the development of a social role. If the total life span may be said to have a cycle, each stage with its unique tasks, then by analogy a role may be said to have a cycle and each stage in that role cycle to have its unique tasks and problems of adjustment. Four broad stages of a role cycle may be specified:

Anticipatory Stage

All major adult roles have a long history of anticipatory training for them, since parental and school socialization of children is dedicated precisely to this task of producing the kind of competent adult valued by the culture. For our present purposes, however, a narrower conception of the anticipatory stage is preferable: the engagement period in the case of the marital role, pregnancy in the case of the parental role, and the last stages of highly vocationally oriented schooling or on-the-job apprenticeship in the case of an occupational role.

Honeymoon Stage

This is the time period immediately following the full assumption of the adult role. The inception of this stage is more easily defined than its termination. In the case of the marital role, the honeymoon stage extends from the marriage ceremony itself through the literal honeymoon and on through an unspecified and individually varying period of time. Raush (Raush, Goodrich, and Campbell 1963) has caught this stage of the marital role in his description of the "psychic honeymoon": that extended postmarital period when, through close intimacy and joint activity, the couple can explore each other's capacities and limitations. I shall arbitrarily consider the onset of pregnancy as marking the end of the honeymoon stage of the marital role. This stage of the parental role may involve an equivalent psychic honeymoon, that post-childbirth period during which, through intimacy and prolonged contact, an attachment between parent and child is laid down. There is a crucial difference, however, from the marital role in this stage. A woman knows her husband as a unique real person when she enters the honeymoon stage of marriage. A good deal of preparatory adjustment on a firm reality base is possible during the engagement period which is not possible in the equivalent pregnancy period. Fantasy is not corrected by the reality of a specific individual child until the birth of the child. The "quickening" is psychologically of special significance to women precisely because it marks the first evidence of a real baby rather than a purely fantasized one. On this basis alone there is greater interpersonal adjustment and learning during the honeymoon stage of the parental role than of the marital role.

Plateau Stage

This is the protracted middle period of a role cycle during which the role is fully exercised. Depending on the specific problem under analysis, one would obviously subdivide this large plateau stage further. For my present purposes it is not necessary to do so, since my focus is on the earlier anticipatory and honeymoon stages of the parental role and the overall impact of parenthood on adults.

Disengagement-Termination Stage

This period immediately precedes and includes the actual termination of the role. Marriage ends with the death of the spouse or,

just as definitively, with separation and divorce. A unique characteristic of parental-role termination is the fact that it is not closely marked
by any specific act but is an attenuated process of termination with
little cultural prescription about when the authority and obligations
of a parent end. Many parents, however, experience the marriage of
the child as a psychological termination of the active parental role.

UNIQUE FEATURES OF THE PARENTAL ROLE

With this role-cycle suggestion as a broader framework, we can narrow
our focus to what are the unique and most salient features of the
parental role. In doing so, special attention will be given to two further
questions: (1) the impact of social changes over the past few decades
in facilitating or complicating the transition to and experience of parenthood and (2) the new interpretations or new research suggested
by the focus on the parent rather than the child.

Cultural Pressure to Assume the Role
        On the level of cultural values, men have no freedom of choice
where work is concerned: They must work to secure their status as
adult men.
        The equivalent for women has been maternity. There is con-
siderable pressure upon the growing girl and young woman to consider maternity necessary for a woman's fulfillment as an individual
and to secure her status as an adult.[1]
        This is not to say there are no fluctuations over time in the
intensity of the cultural pressure to parenthood. During the depression
years of the 1930s, there was more widespread awareness of the
economic hardships parenthood can entail, and many demographic

[1] The greater the cultural pressure to assume a given adult social role, the greater will
be the tendency for individual negative feelings toward that role to be expressed covertly.
Men may complain about a given job, not about working per se, and hence their work
dissatisfactions are often displaced to the nonwork sphere, as psychosomatic complaints
or irritation and dominance at home. An equivalent displacement for women of the
ambivalence many may feel toward maternity is to dissatisfactions with the homemaker
role.

experts believe there was a great increase in illegal abortions during those years. Bird has discussed the dread with which a suspected pregnancy was viewed by many American women in the 1930s (Bird 1966). Quite a different set of pressures were at work during the 1950s, when the general societal tendency was toward withdrawal from active engagement with the issues of the larger society and a turning in to the gratifications of the private sphere of home and family life. Important in the background were the general affluence of the period and the expanded room and ease of child rearing that go with suburban living. During the 1960s there was a drop in the birth rate in general, fourth and higher-order births in particular. During this same period there was increased concern and debate about women's participation in politics and work, with more women returning to work rather than conceiving the third or fourth child.[2]

Inception of the Parental Role

The decision to marry and the choice of a mate are voluntary acts of individuals in our family system. Engagements are therefore consciously considered, freely entered, and freely terminated if increased familiarity decreases, rather than increases, intimacy and commitment to the choice. The inception of a pregnancy, unlike the engagement, is not always a voluntary decision, for it may be the unintended consequence of a sexual act that was recreative in intent rather than procreative. Secondly, and again unlike the engagement, the termination of a pregnancy is not socially sanctioned, as shown by current resistance to abortion-law reform.

The implication of this difference is much higher probability of unwanted pregnancies than of unwanted marriages in our family system. Coupled with the ample clinical evidence of parental rejection and sometimes cruelty to children, it is all the more surprising that there has not been more consistent research attention to the problem of *parental satisfaction*, as there has for long been on *marital satisfaction or work satisfaction*. Only the extreme iceberg tip of the parental satisfaction continuum is clearly demarcated and researched,

---

[2] When it is realized that a mean family size of 3.5 would double the population in forty years, while a mean of 2.5 would yield a stable population in the same period, the social importance of withholding praise for procreative prowess is clear. At the same time, a drop in the birth rate may reduce the number of unwanted babies born, for such a drop would mean more efficient contraceptive usage and a closer correspondence between desired and attained family size.

as in the growing concern with "battered babies." Cultural and psychological resistance to the image of a nonnurturant woman may afflict social scientists as well as the American public.

The timing of a first pregnancy is critical to the manner in which parental responsibilities are joined to the marital relationship. The single most important change over the past few decades is extensive and efficient contraceptive usage, since this has meant for a growing proportion of new marriages, the possibility of an increasing preference for some postponement of childbearing after marriage. When pregnancy was likely to follow shortly after marriage, the major transition point in a woman's life was marriage itself. *This transition point is increasingly the first pregnancy rather than marriage.* It is accepted and increasingly expected that women will work after marriage, while household furnishings are acquired and spouses complete their advanced training or gain a foothold in their work (Davis 1962). This provides an early marriage period in which the fact of a wife's employment presses for a greater egalitarian relationship between husband and wife in decision-making, commonality of experience, and sharing of household responsibilities.

The balance between individual autonomy and couple mutuality that develops during the honeymoon stage of such a marriage may be important in establishing a pattern that will later affect the quality of the parent-child relationship and the extent of sex-role segregation of duties between the parents. It is only in the context of a growing egalitarian base to the marital relationship that one could find, as Gavron (1966) has, a tendency for parents to establish some barriers between themselves and their children, a marital defense against the institution of parenthood as she describes it. This may eventually replace the typical coalition in more traditional families of mother and children against husband-father. Parenthood will continue for some time to impose a degree of temporary segregation of primary responsibilities between husband and wife, but, when this takes place in the context of a previously established egalitarian relationship between the husband and wife, such role segregation may become blurred, with greater recognition of the wife's need for autonomy and the husband's role in the routines of home and child rearing. . . .

Irrevocability

If marriages do not work out, there is now widespread acceptance of divorce and remarriage as a solution. The same point applies to the work world: We are free to leave an unsatisfactory job and seek another. But once a pregnancy occurs, there is little possibility of undoing the commitment to parenthood implicit in conception except in the rare instance of placing children for adoption. We can have ex-spouses and ex-jobs but not ex-children. This being so, it is scarcely surprising to find marked differences between the relationship of a parent and one child and the relationship of the same parent with another child. If the culture does not permit pregnancy termination, the equivalent to giving up a child is psychological withdrawal on the part of the parent. . . .

Mention was made earlier that for many women the personal outcome of experience in the parent role is not a higher level of maturation but the negative outcome of a depressed sense of self-worth, if not actual personality deterioration. There is considerable evidence that this is more prevalent than we recognize. On a qualitative level, a close reading of the portrait of the working-class wife in Rainwater (1959), Newsom (1963), Komarovsky (1962), Gavron (1966), or Zweig (1952) gives little suggestion that maternity has provided these women with opportunities for personal growth and development. So, too, Cohen (1966) notes with some surprise that in her sample of middle-class educated couples, as in Pavenstadt's study of lower-income women in Boston, there were more emotional difficulties and lower levels of maturation among multiparous women than primiparous women. On a more extensive sample basis, in Gurin's (1960) survey of Americans viewing their mental health, as in Bradburn's (1965) reports on happiness, single men are less happy and less active than single women, but among the married respondents the women are unhappier, have more problems, feel inadequate as parents, have a more negative and passive outlook on life, and show a more negative self-image. All of these characteristics increase with age among married women but show no relationship to age among men. While it may be true, as Gurin argues, that women are more introspective and hence more attuned to the psychological facets of experience than men are, this point does not account for the fact that the things which the women report are all on the negative side; few are on the positive side, indicative of euphoric sensitivity and pleasure. The possibility must be faced, and at some point researched, that women lose ground in personal development and self-esteem during the early and middle years of adulthood, whereas men gain ground in these respects during the same years. The retention of a high level of self-esteem may depend upon the adequacy of earlier preparation for major adult roles: Men's training adequately prepares them for their primary adult roles in the occupational system, as it does for those women who opt to participate significantly in the work world. Training in the qualities and skills needed for family roles in contemporary society may be inadequate for both sexes, but the lowering of self-esteem occurs only among women because their primary adult roles are within the family system.

Preparation for Parenthood

Four factors may be given special attention on the question of what preparation American couples bring to parenthood.

Paucity of Preparation. Our educational system is dedicated to the cognitive development of the young, and our primary teaching approach is the pragmatic one of learning by doing. How much one knows and how well he can apply what he knows are the standards by which the child is judged in school, as the employee is judged at work. The child can learn by doing in such subjects as science, mathematics, artwork, or shop, but not in the subjects most relevant to successful family life: sex, home maintenance, child care, interpersonal competence, and empathy. If the home is deficient in training in these areas, the child is left with no preparation for a major segment of his adult life. A doctor facing his first patient in private practice has treated numerous patients under close supervision during his internship, but probably a majority of American mothers approach maternity with no previous child-care experience beyond sporadic baby-sitting, perhaps a course in child psychology, or occasional care of younger siblings.

Limited Learning during Pregnancy. A second important point makes adjustment to parenthood potentially more stressful than marital adjustment. This is the lack of any realistic training for parenthood during the anticipatory stage of pregnancy. By contrast, during the engagement period preceding marriage, an individual has opportunities to develop the skills and make the adjustments which ease the transition to marriage. Through discussions of values and life goals, through sexual experimentation, shared social experiences as an engaged couple with friends and relatives, and planning and furnishing an apartment, the engaged couple can make considerable progress in developing mutuality in advance of the marriage itself (Rapoport 1963; Raush et al. 1963). No such headstart is possible in the case of pregnancy. What preparation exists is confined to reading, consultation with friends and parents, discussions between husband and wife, and a minor nesting phase in which a place and the equipment for a baby are prepared in the household.[3]

Abruptness of Transition. Third, the birth of a child is not followed by any gradual taking on the responsibility, as in the case of a professional work role. It is as if the woman shifted from a graduate student to a full professor with little intervening apprenticeship experience of slowly increasing responsibility. The new mother starts our immediately on 24-hour duty, with responsibility for a fragile and mysterious infant totally dependent on her care.

   If marital adjustment is more difficult for very young brides than more mature ones (Burchinal 1959; Martinson 1955; Moss and Gingles 1959), adjustment to motherhood may be even more difficult. A woman can adopt a passive dependence on a husband and still have a successful marriage, but a young mother with strong dependency needs is in for difficulty in maternal adjustment, because the role precludes such dependency. This situation was well described in Cohen's (1966) study in a case of a young wife with a background of coed popularity and a passive dependent relationship to her admired and admiring husband, who collapsed into restricted incapacity when faced with the responsibilities of maintaining a home and caring for a child.

Lack of Guidelines to Successful Parenthood. If the central task of parenthood is the rearing of children to become the kind of competent adults valued by the society, then an important question facing any parent is what he or she specifically can do to create such a competent adult. This is where the parent is left with few or no guidelines from the expert. Parents can readily inform themselves concerning the

---

[3] During the period when marriage was the critical transition in the adult woman's life rather than pregnancy, a good deal of anticipatory ''nesting'' behavior took place from the time of conception. Now more women work through a considerable portion of the first pregnancy, and such nesting behavior as exists may be confined to a few shopping expeditions or baby showers, thus adding to the abruptness of the transition and the difficulty of adjustment following the birth of a first child.

young infant's nutritional, clothing, and medical needs and follow the general prescription that a child needs loving physical contact and emotional support. Such advice may be sufficient to produce a healthy, happy, and well-adjusted preschooler, but adult competency is quite another matter. . . .

Brim (1957) points out that we are a long way from being able to say just what parent-role prescriptions have what effect on the adult characteristics of the child. We know even less about how such parental prescriptions should be changed to adapt to changed conceptions of competency in adulthood. In such an ambiguous context, the great interest parents take in school reports on their children or the pediatrician's assessment of the child's developmental progress should be seen as among the few indices parents have of how well *they* are doing as parents. . . .

REFERENCES

Benedek, T. 1959. Parenthood as a developmental phase. *J. American Psychoanalytic Assoc.* 7:389–417.

Bird, C. 1966. *The Invisible Scar.* New York: McKay.

Bradburn, N. and D. Caplovitz. 1965. *Reports on Happiness.* Chicago: Aldine.

Brim, O. 1957. The parent-child relation as a social system: I. Parent and child roles. *Child Development* 28:343–364.

Burchinal, L. 1959. Adolescent role deprivation and high school marriage. *Marriage and Family Living* 21:378–384.

Cohen, M. 1966. Personal identity and sexual identity. *Psychiatry* 29:1–14.

Davis, J. 1962. *Stipends and Spouses: The Finances of American Arts and Sciences Graduate Students.* Chicago: University of Chicago Press.

Gavron, H. 1966. *The Captive Wife.* London: Routledge and Kegan Paul.

Gurin, G., J. Veroff, and S. Feld. 1960. *Americans View Their Mental Health.* New York: Basic Books.

Lidz, T., S. Fleck, and A. Cornelison. 1965. *Schizophrenia and the Family.* New York: International Universities Press.

Komarovsky, M. 1962. *Blue Collar Marriage.* New York, Random House.

Moss, J. and R. Gingles 1959. The relationship of personality to the incidence of early marriage. *Marriage and Family Living* 21:373–377.

Newsom, J. and E. Newsom. 1963. *Infant Care in an Urban Community.* New York: International Universities Press.

Parsons, T. and R. Bales. 1955. *Family Socialization and Interaction Process.* New York: Free Press.

Rainwater, L., R. Coleman, and G. Havidel. 1959. *Working Man's Wife.* New York: Oceana Publications.

Rapoport, R. 1963. Normal crises, family structure, and mental health. *Family Process* 2:68–80.

Raush, H., W. Goodrich and J. Campbell. 1963. Adaptation to the first years of marriage. *Psychiatry* 26:368–380.

Rheingold, J. 1964. *The Fear of Being a Woman: A Theory of Maternal Destructiveness.* New York: Grune and Stratton.

Yarrow, L. 1961. Maternal deprivation: toward an empirical and conceptual reevaluation. *Psychological Bulletin* 58:459–490.

Zweig, F. 1952. *Woman's Life and Labor.* London: Camelot Press.

Twenty

Work and Its Meaning
Lillian Breslow Rubin

How important is work for the continuing development of a sense of mastery and competence during the adult years? The answer to this question depends largely on the type of job the individual holds. In this selection, Lillian Rubin presents excerpts from extensive interviews she has conducted with working-class men and women. For many of the individuals interviewed in this study, work is unchallenging, dissatisfying, and not particularly meaningful. A crucial determinant of whether an individual derives meaning from work appears to be whether the work permits the individual to excercise freedom, autonomy, and competence.

Tell me something about the work you do and how you feel about it.

FOR THE MEN, whose definition of self is so closely tied to work, it's a mixed bag—a complex picture of struggle, of achievements and disappointments, of successes and failures. In their early work life, most move restlessly from job to job seeking not only higher wages and better working conditions, but some kind of work in which they can find meaning, purpose, and dignity:

> God, I hated that assembly line. *I hated it.* I used to fall asleep on the job standing up and still keep doing my work. There's nothing more boring and more repetitious in the world. On top of it, you don't feel human. The machine's running you, you're not running it.
>
> *(33-year-old mechanic)*

Thus, by the time they're twenty-five, their post-school work life averages almost eight years, and half have held as many as six, eight, or ten jobs.

255

Generally, they start out as laborers, operatives in an oil refin-
ery, assembly-line workers in the local canneries, automobile or parts
plants, warehousemen, janitors, or gas station attendants—jobs in which
worker dissatisfaction is well documented. Some move on the up—into
jobs that require more skill, jobs that still demand plenty of hard work,
but which at least leave one with a sense of mastery and competence:

> I'm proud of what I've done with my life. I come from humble
> origins, and I never even finished school; but I've gotten some-
> place. I work hard, but it's good work. It's challenging and never
> routine. When I finish a day's work, I know I've accomplished
> something. I'm damned good at what I do, too. Even the boss
> knows it.
>
> (*36-year-old steam fitter*)

But the reality of the modern work world is that there are fewer
and fewer jobs calling for such traditional skills. So most job changes
don't mean moving up, but only moving on:

> When I first started, I kept moving around. I kept looking for a
> job I'd like. You know, a job where it wouldn't make you tired
> just to get up in the morning and have to go to work. [*With a
> heavy sigh.*] It took me a number of years to discover that there's
> not much difference—a job's a job. So now I do what I have to do,
> and maybe I can get my family a little security.
>
> (*27-year-old mail sorter*)

For some, the job changes are involuntary—due to layoffs:

> When I first got out of high school, I had a series of jobs and
> a series of layoffs. The jobs lasted from three weeks to three
> months. Something always happened—like maybe the contract
> didn't come through—and since I was low man on the totem pole,
> I got laid off. A lot of times, the layoffs lasted longer than the
> jobs.

. . . or industrial accidents—a common experience among men who
work in factories, warehouses, and on construction sites:

> I was working at the cannery about a week when my hand got
> caught in the belt. It got crushed, and I couldn't work for three

months. When I got better, they wouldn't put me back on the job because they said I was accident-prone.

By the time they're thirty, about half are settled into jobs at which they've worked for five years. With luck, they'll stay at them for many more to come. Without it, like their fathers, they'll know the pain of periodic unemployment, the fear of their families doing without. For the other half—those still floating from job to job—the future may be even more problematic. Unprotected by seniority, with work histories that prospective employers are likely to view as chaotic and unstable, they can expect little security from the fluctuations and uncertainties of the labor market.

But all that tells us nothing about the quality of life on the job— what it feels like to go to work at *that* particular job for most of a lifetime—an experience that varies in blue-collar jobs just as it does in white-collar ones. For just as there are elite jobs in the white-collar work force, so they exist among blue-collar workers. Work that allows for freedom and autonomy on the job—these are the valued and high-status jobs, rare in either world. For the blue-collar worker, that means a job where he can combine skill with strength, where he can control the pace of his work and the order of the tasks to be done, and where successful performance requires his independent judgments. To working-class men holding such jobs—skilled construction workers, skilled mechanics, truck drivers—the world of work brings not only goods, but gratifications. The man who drives the long-distance rig feels like a free agent once he gets out on the road. It's true, there's a time recorder on the truck that clocks his stops. Still, compared to jobs he's had before in factories and warehouses, on this one, he's the guy who's in control. Sometimes the road's easy; sometimes it's tough. Always it requires his strength and skill; always he's master of the machine:

> There's a good feeling when I'm out there on the road. There ain't nobody looking over your shoulder and watching what you're doing. When I worked in a warehouse, you'd be punching in and punching out, and bells ringing all the time. On those jobs, you're not thinking, you're just doing what they tell you. Sure, now I'm expected to bring her in on time, but a couple of hours one way or the other don't make no difference. And there ain't nobody but me to worry about how I get her there.
>
> *(28-year-old trucker)*

The skilled construction worker, too, finds challenge and reward in his work:

> I climb up on those beams every morning I'm working, and I like being way up there looking down at the world. It's a challenge up there, and the work's hardly ever routine. You have to pay attention and use your head, too, otherwise you can get into plenty of trouble in the kind of work I do. I'm a good man, and everybody on the job knows it.
>
> *(31-year-old ironworker)*

But most blue-collar men work at jobs that require less skill, that have less room for independent judgment—indeed, often expect that it will be suspended—and that leave their occupants with little freedom or autonomy. Such jobs have few intrinsic rewards and little status—either in the blue-collar world or the one outside—and offer few possibilities for experiencing oneself as a "good man." The men who hold these jobs often get through each day by doing their work and numbing themselves to the painful feelings of discontent—trying hard to avoid the question, "Is this what life is all about?" Unsuccessful in that struggle, one 29-year-old warehouseman burst out bitterly:

> A lot of times I hate to go down there. I'm cooped up and hemmed in. I feel like I'm enclosed in a building forty hours a week, sometimes more. It seems like all there is to life is to go down there and work, collect your paycheck, pay your bills, and get further in debt. It doesn't seem like the circle ever ends. Every day it's the same thing; every week it's the same thing; every month it's the same thing.

Some others respond with resignation:

> I guess you can't complain. You have to work to make a living, so what's the use.
>
> *(26-year-old garage man)*

. . . some with boredom:

> I've been in this business thirteen years and it bores me. It's enough.
>
> *(35-year-old machine operator)*

. . . some with alienation:

> The one thing I like is the hours. I work from seven to three-thirty
> in the afternoon so I get off early enough to have a lot of the day
> left.
>
> (*28-year-old assembly-line worker*)

All, in fact, probably feel some combination of all these feelings. For
the men in such jobs, bitterness, alienation, resignation, and boredom
are the defining features of the work experience. For them, work is
something to do, not to talk about. "What's there to talk about?"—not
really a question but an oft-repeated statement that says work is a
requirement of life, hours to be gotten through until you can go
home. . . .

Under the pressures of these financial strains, 58 percent of
the working-class wives work outside the home—most in part-time jobs.
Of those who stay at home, about two-thirds are happy to do so,
considering the occupation of "homemaker and mother" an important
and gratifying job. Some are glad to work only in the home because
jobs held earlier were experienced as dull and oppressive:

> I worked as a file clerk for Montgomery Ward's. I hated it. There
> was always somebody looking over your shoulder trying to catch
> you in mistakes. Besides, it was boring; you did the same thing
> all day long. Now I can stop when I don't feel like doing some-
> thing and play with the children. We go for walks, or we work
> in the garden.

. . . some, because life outside seems frightening:

> No, I don't ever want to work again if I don't have to. It's really
> too hectic out there. Now when I'm home, I can go out to it when
> I want. I suppose it sounds like I'm hiding from something, or
> escaping from it. But I'm not. It's just that sometimes it's
> overwhelming.

. . . and some, just because they enjoy both the tasks and the freedom
of work in the home:

> I wouldn't work ever again if I didn't have to. I like staying home.
> I sew and take care of the house and kids. I go shopping. I'm

my own boss. I like that. And I also like fixing up the house and making it look real nice. And I like cooking nice meals so Ralph is proud of me.

But few working-class wives are free to make the choice about working inside or outside the home depending only on their own desires. Most often, economic pressures dictate what they will do, and *even those who wish least to work outside the home probably will do so sometime in their lives.* Thus, for any given family, the wife is likely to move in and out of the labor force depending on the husband's job stability, on whether his overtime expands or contracts, on the exigencies of family life—a sick child, an aging parent.

The women I met work as beauticians, sales clerks, seam-stresses, cashiers, waitresses, office clerks, typists, occasionally as secretaries and factory workers; and at a variety of odd jobs such as baby-sitters, school-crossing guards, and the like. Their work hours range from a few hours a week to a few—nine in all—who work full time. Most—about three-quarters—work three or four days a week regularly.

Their attitudes toward their work are varied, but most find the work world a satisfying place—at least when compared to the world of the homemaker. Therefore, although many of these women are pushed into the job market by economic necessity, they often stay in it for a variety of other reasons.

An anomaly, a reader might say. After all, hasn't it already been said that wives who hold jobs outside the home often are resentful because they also bear most of the burden of the work inside the home? Yet both are true. Women can feel angry and resentful because they are overburdened when trying to do both jobs almost single-handedly, while at the same time feeling that work outside the home provides satisfactions not otherwise available. Like men, they take pride in doing a good job, in feeling competent. They are glad to get some relief from the routines of homemaking and mothering small children. They are pleased to earn some money, to feel more inde-pendent, more as if they have some ability to control their own lives. Thus, they ask no more—indeed, a good deal less—than men do; the chance to do work that brings such rewards while at the same time having someone to share some of the burdens of home and family.

There is, perhaps, no greater testimony to the deadening and deadly quality of the tasks of the homemaker than the fact that so many women find pleasure in working at jobs that by almost any definition would be called alienated labor—low-status, low-paying, dead-end work made up of dull, routine tasks; work that often is considered too menial for men who are less educated than these women. Nor is there greater testimony to the efficacy of the socialization process. Bored and discontented with the never ending routine of household work, they seek stimulation in work outside the home. But a lifetime of preparation for homemaking and motherhood makes it possible to find gratification in jobs that require the same qualities—service, submission, and the suppression of intellectual development.

No accident either that these traits are the ideal complements for the needs of the economy for a cheap, supplemental labor pool that can be moved in and out of the labor force as the economy expands and contracts. Indeed, the sex-stereotyped family roles dovetail neatly with this requirement of our industrial economy. With each expansion, women are recruited into the labor force at the lowest levels. Because they are defined primarily in their family roles rather than as workers, they are glad to get whatever work is available. For the same reason, they are willing to work for wages considerably below those of men. When the economy contracts, women are expected to give up their jobs and to return quietly to the tasks of homemaking and mothering. Should they resist, they are reminded with all the force that society can muster that they are derelict in their primary duties and that those they love most dearly will pay a heavy price for their selfishness.

Tell me something about the work you do and how you feel about it.

A thirty-one-year-old factory worker, mother of five children, replies:

> I really love going to work. I guess it's because it gets me away from home. It's not that I don't love my home; I do. But you get awfully tired of just keeping house and doing those housewifely things. Right now, I'm not working because I was laid off last month. I'm enjoying the layoff because things get awfully hectic at work, but it's only a short time. I wouldn't like to be off for a long time. Anyhow, even now I'm not completely not working. I've been waiting tables at a coffee shop downtown. I like the people down there, and it's better than not doing anything.

> You know, when I was home, I was getting in real trouble. I had that old housewife's syndrome, where you either crawl in bed after the kids go to school or sit and watch TV by the hour. I

was just dying of boredom and the more bored I got, the less and less I did. On top of it, I was getting fatter and fatter, too. I finally knew I had to do something about it, so I took this course in upholstery and got this job as an upholstery trimmer.

"It gets me away from home"—a major reason why working women of any class say they would continue to work even apart from financial necessity. For most, however, these feelings of wanting to flee from the boredom and drudgery of homemaking are held ambivalently as they struggle with their guilt about leaving young children in someone else's care.

For all women, the issues around being a "working mother" are complex, but there are some special ones among the working class that make it both harder and easier for women to leave their homes to work. It is harder because, historically, it has been a source of status in working-class communities for a woman to be able to say, "I don't *have* to work." Many men and women still feel keenly that it's his job to support the family, hers to stay home and take care of it. For her to take a job outside the home would be, for such a family, tantamount to a public acknowledgement of his failure. Where such attitudes are still held strongly, sometimes the wife doesn't work even when it's necessary; sometimes she does. Either way, the choice is difficult and painful for both.

On the other hand, it's easier for the wives of working-class men to override their guilt about leaving the children because the financial necessity is often compelling. On one level, that economic reality is an unpleasant one. On another, it provides the sanction for leaving the home and makes it easier for working-class women to free themselves from the inner voices that charge, "You're self-indulgent," that cry, "What kind of mother are you?"—as this conversation with a twenty-five-year-old working mother of two shows:

How do you feel about working?

I enjoy it. It's good to get out of the house. Of course, I wouldn't want to work full time; that would be being away from the kids too much.

Do you sometimes wish you could stay home with them more?

Yeah, I do.

What do you think your life would look like if you could?

> Actually, I don't know. I guess I'd get kind of bored. I don't mean that I don't enjoy the kids; I do. But you know what I mean. It's kind of boring being with them day after day. Sometimes I feel bad because I feel like that. It's like my mind battles with itself all the time—like, "Stay home" and "Go to work."

So you feel guilty because you want to work and, at the same time, you feel like it would be hard for you to stay home all the time?

> Yeah, that's right. Does it sound crazy?

No, it doesn't. A lot of women feel that way. I remember feeling that way when my children were young.

> You, too, huh? That's interesting. What did you do?.

Sometimes I went to work, and sometimes I stayed home. That's the way a lot of women resolve that conflict. Do you think you'd keep on working even if you didn't need the money at all?

> I think about that because Ed says I could stop now. He says we can make it on his salary and that he wants me home with the kids. I keep saying no, because we still need this or that. That's true, too. It would be really hard. I'm not so sure we could do it without my salary. Sometimes I think he's not sure either. I've got to admit it, though, I don't really want to stay home. I wouldn't mind working three days instead of four, but that's about all. I guess I really work because I enjoy it. I'm good at it, and I like that feeling. It's good to feel like you're competent.

So you find some real gratifications in your work. Do you also sometimes think life would be easier if you didn't work?

> Sure, in some ways, but maybe not in others. Anyhow, who expects life to be easy? Maybe when I was a kid I thought about things like that, but not now.

Faced with such restlessness, women of any class live in a kind of unsteady oscillation between working and not working outside the home—each choice exacting its own costs, each conferring its own rewards. Another woman, thirty-two and with four children, chooses differently, at least for now:

> Working is hard for me. When I work, I feel like I want to be doing a real good job, and I want to be absolutely responsible. Then, if the little one gets a cold, I feel like I should be home with her. That causes complications because whatever you do isn't right. You're either at work feeling like you should be home with your sick child, or you're at home feeling like you should be at work.

So right now, you're relieved at not having to go to work?

> Yeah, but I miss it, too. The days go faster and they're more exciting when you work.

Do you think you'll go back to work, then?

> Right now, we're sort of keeping up with the bills, so I probably won't. When we get behind a lot again, I guess I'll go back then.

Thus, the "work–not work" issue is a lively and complicated one for women—one whose consequences radiate throughout the marriage and around which important issues for both the individuals and the marital couple get played out. Even on the question of economic necessity, wives and husbands disagree in a significant minority of the families. For "necessity" is often a relative term, the definition ultimately resting on differences between wives and husbands on issues of value, life-style, sex-role definitions, and conjugal power. Thus, he says:

> She doesn't have to work. We can get by. Maybe we'll have to take it easy on spending, but that's okay with me. It's worth it to have her home where she belongs.

She says:

> My husband says I don't have to work, but if I don't, we'll never get anywhere. I guess it's a matter of pride with him. It makes

him feel bad, like he's not supporting us good enough. I understand how he feels, but I also know that, no matter what he says, if I stop working, when the taxes on the house have be paid, there wouldn't be any money if we didn't have my salary.

In fact, both are true. The family *could* lower its living standard—live in an apartment instead of a house; have less, do less. On his income of about $11,500, they undoubtedly could survive. But with all his brave words about not wanting his wife to work, he is not without ambivalence about the consequences. He is neither eager to give up the few comforts her salary supports nor to do what he'd have to do in order to try to maintain them. She says:

He talks about me not working, then right after I went back this time, he bought this big car. So now, I have to work or else who would make the payments?

He says:

If she stops working, I'd just get a second job so we could keep up this place and all the bills and stuff.

How do you feel about having to do that?

Well, I wouldn't exactly love it. Working two jobs with hardly any time off for yourself isn't my idea of how to enjoy life. But if I had to, I'd do it.

What about the payments on the car? Wouldn't they get to be a big problem if she didn't work?

Yeah, that's what she says. I guess she's right. I don't want her to work; but even if I worked at night, too, I don't know how much I could make. She's right about if I work two jobs then I wouldn't have time to do anything with the family and see the kids. That's no life for any of us, I guess.

The choices, then, for this family, as for so many others, are difficult and often emotionally costly. In a society where people in all classes are trapped in frenetic striving to acquire goods, where a man's sense of worth and his definition of his manhood rest heavily on his ability to provide those goods, it is difficult for him to acknowledge that the family really does need his wife's income to live as they both would like. Yet, just beneath the surface of his denial is understanding—understanding that he sometimes experiences with pain, sometimes masks with anger. His wife understands his feelings. "It's a matter of pride with him," she says. "It makes him feel bad, like he's not supporting us good enough," she says. But she also knows that he, like she, wants the things her earnings buy.

It should be clear by now that for most women there are compensations in working outside the home that go beyond the material ones—a sense of being a useful and valued member of society:

> If you don't bring home a paycheck, there's no gauge for whether you're a success or not a success. People pay you to work because you're doing something useful and you're good at it. But nobody pays a housewife because what difference does it make; nobody really cares.
>
> (*34-year-old typist*)

. . . of being competent:

> In my work at the salon, it's really like an ego trip. It feels good when people won't come in if you're not there. If I go away for two weeks, my customers will wait to have their hair done until I come back. I'm not always very secure. but when I think about that, it always makes me feel good about myself, like I'm really okay.
>
> (*31-year-old beautician*)

. . . of feeling important:

> I meet all kinds of interesting people at work, and they depend on me to keep the place nice. When I don't go in sometimes, the place gets to be a mess. Nobody sweeps up, and sometimes they don't even call to have a machine fixed. It makes me feel good—you know, important—when I come back and everybody is glad to see me becuase they know everything will be nice again.
> (*29-year-old manager of a self-service laundromat*)

. . . and of gaining a small measure of independence from their husbands:

> I can't imagine not working. I like to get out of the house, and the money makes me feel more independent. Some men are funny. They think if you don't work, you ought to just be home every day, like a drudge around the house, and that they can come home and just say, "Do this," and "Do that," and "Why is that dish in the sink?" When you work and make some money, it's different. It makes me feel more equal to him. He can't just tell me what to do.

In fact, students of the family have produced a large literature on intrafamily power which shows that women who work outside the house have more power inside the house. Most of these studies rest on the resource theory of marital power—a theory which uses the language of economics to explain marital relations. Simply stated, resource theory conceptualizes marriage as a set of exchange relations in which the balance of power will be on the side of the partner who contributes the greater resources to the marriage. While not made explicit, the underlying assumption of this theory is that the material contributions of the husband are the "greater resource." The corollary, of course, is the implicit denigration and degradation of the functions which women traditionally perform in the household—not the least of them providing the life-support system, the comfort, and the respite from the outside world that enables men to go back into it each day.

So pervasive is the assumption of the greater importance of the male contribution to the family that generations of social scientists have unthinkingly organized their research around this thesis. Unfortunately, however, it is not the social scientists alone who hold this view. For women as well too often accept these definitions of the value of their role in the family and do, in fact, feel more useful, more independent, more able to hold their own in a marital conflict when they are also working outside the home and contributing some share of the family income. Such is the impact of the social construction of reality; for, as the old sociological axiom says: "If men define situations as real, they are real in their consequences."

Indeed, it is just this issue of her independence that is a source of conflict in some of the marriages where women work. Mostly, when women hold outside jobs, there is some sense of partnership in a joint enterprise—a sharing of the experience of two people working together for a common goal. But in well over one-third of the families, husbands complain that their working wives "are getting too independent." Listen to this conversation with a thirty-three-year-old repairman:

> She just doesn't know how to be a real wife, you know, feminine and really womanly. She doesn't know how to give respect because she's too independent. She feels that she's a working woman and she puts in almost as many hours as I do and brings home a paycheck, so there's no one person above the other. She doesn't want there to be a king in this household.

And you want to be a king?

> No, I guess I don't really want to be a king. Well [*laughing*], who wouldn't want to be? But I know better. I just want to be recognized as an important individual. She needs to be more feminine. When she's able to come off more feminine than she is, then maybe we'll have something deeper in this marriage.

I'm not sure I know what you mean. Could you help me to understand what you want of her?

> Look, I believe every woman has the right to be an individual, but I just don't believe in it when it comes between two people. A man needs a feminine woman. When it comes to two people living together, a man is supposed to be a man and a woman is supposed to be a woman.

But just what does that mean to you?

> I'd like to feel like I wear the pants in the family. Once my decision is made, it should be made, and that's it. She should just carry it out. But it doesn't work that way around here. Because she's working and making money, she thinks she can argue back whenever she feels like it.

Another man, one who has held eight jobs in his seven-year marriage, speaks angrily:

> I think our biggest problem is her working. She started working and she started getting too independent. I never did want her

to go to work, but she did anyway. I don't think I had the say-so that I should have.

It sounds as if you're feeling very much as if your authority has been challenged on this issue of her working.

You're damn right. I feel the man should do the work, and he should bring home the money. And when he's over working, he should sit down and rest for the rest of the day.

And you don't get to do that when she's working?

Yeah, I do it. But she's got a big mouth so it's always a big hassle and fight. I should have put my foot down a long time ago and forced her into doing things my way.

The women respond to these charges angrily and defensively. The men are saying: be dependent, submissive, subordinate—mandates with which all women are reared. But for most white working-class women—as for many of their black sisters—there is a sharp disjunction between the commandments of the culture and the imperatives of their experience.

The luxury of being able to depend on someone else is not to be theirs. And often, they are as angry at their men for letting them down as the men are at the women for not playing out their roles in the culturally approved ways. A thirty-two-year-old mother of two speaks:

I wish I could be dependent on him like he says. But how can you depend on someone who does the things he does? He quits a job just because he gets mad. Or he does some dumb thing, so he gets fired. If I didn't work, we wouldn't pay the rent, no matter what he says.

Another thirty-year-old mother of three says:

He complains that I don't trust him. Sure I don't. When I was pregnant last time and couldn't work, he went out with his friends and blew money around. I never know what he's going to do. By the time the baby came, we were broke, and I had to go back to work before she was three weeks old. It was that or welfare. Then he complains because I'm too independent. Where would we be if I wasn't?

Thus are both women and men stuck in a painful bind, each blaming the other for failures to meet cultural fantasies—fantasies that have little relation to their needs, their experiences, or the socioeconomic realities of the world they live in. She isn't the dependent, helpless, frivolous child-woman because it would be ludicrously inappropriate, given her life experiences. He isn't the independent, masterful, all-powerful provider, not because he does "dumb" or irresponsible things, but because the burdens he carries are too great for all but a few of the most privileged—burdens that are especially difficult to bear in a highly competitive economic system that doesn't grant every man and woman the right to work at a self-supporting and self-respecting wage as a matter of course. . . .

Adulthood
Identity and the Self

Twenty-One

The Subjective Experience of Middle Age

Bernice L. Neugarten and Nancy Datan

How do middle-aged adults see themselves? In this selection on identity and self during adulthood, Neugarten and Datan discuss the experience of midlife from the point of view of the adult. The authors suggest that this point in the life cycle is a time of increased self-awareness, introspection, and heightened self-understanding. Compared with the adolescent, the midlife adult is much more cognizant of his or her changing sense of self.

MIDDLE-AGED MEN AND WOMEN, while they recognize the rapidity of social change and while they by no means regard themselves as being in command of all they survey, nevertheless recognize that they constitute the powerful age group vis-à-vis other age groups; that they are the norm bearers and the decision-makers; and that they live in a society that, while it may be oriented toward youth, is controlled by the middle-aged. There is space here to describe only a few of the psychological issues of middle age as they have been described in one of our studies in which 100 highly placed men and women were interviewed at length concerning the salient characteristics of middle adulthood (Neugarten 1967). These people were selected randomly from various directories of business leaders, professionals, and scientists.

The enthusiasm manifested by these informants as the interviews progressed was only one of many confirmations that middle age is a period of heightened sensitivity to one's position within a complex social environment and that reassessment of the self is a prevailing theme. As anticipated, most of this group were highly introspective and verbal persons who evidenced considerable insight into the changes that had taken place in their careers, their families,

273

their status, and in the ways in which they dealt with both their inner and outer worlds. Generally the higher the individual's career position, the greater was his willingness to explore the various issues and themes of middle age.

THE DELINEATION OF MIDDLE AGE

There is ample evidence in these reports, as in the studies mentioned earlier, that middle age is perceived as a distinctive period, one that is qualitatively different from other age periods. Middle-aged people look to their positions within different life contexts—the body, the career, the family—rather than to chronological age for their primary cues in clocking themselves. Often there is a differential rhythm in the timing of events within these various contexts, so that the cues utilized for placing oneself in this period of the life cycle are not always synchronous. For example, one business executive regards himself as being on top in his occupation and assumes all the prerogatives that go with seniority in that context, yet, because his children are still young, he feels he has a long way to go before completing his major goals within the family.

DISTANCE FROM THE YOUNG

Generally the middle-aged person sees himself as the bridge between the generations, both within the family and within the wider contexts of work and community. At the same time he has a clear sense of differentiation from both the younger and older generations. In his view young people cannot understand or relate to the middle-aged because they have not accumulated the prerequisite life experiences. . . . The middle-ager becomes increasingly aware of the distance—emotionally, socially, and culturally—between himself and the young Sometimes the awareness comes as a sudden revelation:

> I used to think that all of us in the office were contemporaries, for we all had similar career interests. But one day we were

talking about old movies and the younger ones had never seen a Shirley Temple film. . . . Then it struck me with a blow that I was older than they. I had never been so conscious of it before.

Similarly, another man remarked:

When I see a pretty girl on the stage or in the movies, and when I realize she's about the age of my son, it's a real shock. It makes me realize that I'm middle-aged.

An often expressed preoccupation is how one should relate to both younger and older persons and how to act one's age. Most of our respondents are acutely aware of their responsibility to the younger generation and of what we called "the creation of social heirs." One corporation executive says,

I worry lest I no longer have the rapport with young people that I had some years back. I think I'm becoming uncomfortable with them because they're so uncomfortable with me. They treat me like I treated my own employer when I was twenty-five. I was frightened of him. . . . But one of my main problems now is to encourage young people to develop so that they'll be able to carry on after us.

And a fifty-year-old-woman says,

You always have younger people looking to you and asking questions. . . . You don't want them to think you're a fool. . . . You try to be an adequate model.

The awareness that one's parents' generation is now quite old does not lead to the same feeling of distance from the parental generation as from the younger generation.

I sympathize with old people, now, in a way that is new. I watch my parents, for instance, and I wonder if I will age in the same way.

The sense of proximity and identification with the old is enhanced by the feeling that those who are older are in a position to understand

and appreciate the responsibilities and commitments to which the middle-aged have fallen heir.

> My parents, even though they are much older, can understand what we are going through; just as I now understand what they went through.

Although the idiosyncrasies of the aged may be annoying to the middle-aged, an effort is usually made to keep such feelings under control. There is greater projection of the self in one's behavior with older people, sometimes to the extent of blurring the differences between the two generations. One woman recounted an incident that betrayed her apparent lack of awareness (or her denial) of her mother's aging:

> I was shopping with mother. She had left something behind on the counter and the clerk called out to tell me that the "old lady" had forgotten her package. I was amazed. Of course, the clerk was a young man and she must have seemed old to him. But I myself don't think of her as old.

MARRIAGE AND FAMILY

Women, but not men, tend to define their age status in terms of the timing of events within the family cycle. Even unmarried career women often discuss middle age in terms of the family they might have had.

> Before I was thirty-five, the future just stretched forth, far away. . . . I think I'm doing now what I want. The things that troubled me in my thirties about marriage and children don't bother me now because I'm at the age where many women have already lost their husbands.

Both men and women, however, recognized a difference in the marriage relationship that follows upon the departure of children, some describing it in positive, others in negative terms, but all recognizing a new marital adjustment to be made.

It's a totally new thing. Now there isn't the responsibility for the children. There's more privacy and freedom to be yourself. All of a sudden there are times when we can just sit down and have a conversation. And it was a treat to go on a vacation alone!

It's the boredom that has grown up between us but which we didn't face before. With the kids at home, we found something to talk about, but now the buffer between us is gone. There are just the two of us, face to face. . . .

One difference between husbands and wives is marked. Most of the women interviewed feel that the most conspicuous characteristic of middle age is the sense of increased freedom. Not only is there increased time and energy available for the self, but there is also a satisfying change in self-concept. The typical theme is that middle age marks the beginning of a period in which latent talents and capacities can be put to use in new directions.
Some of these women describe this sense of freedom coming at the same time that their husbands are reporting increased job pressures or—something equally troublesome—job boredom. Contrast this typical statement of a woman.

I discovered these last few years that I was old enough to admit to myself the things I could do well and to start doing them. I didn't think like this before. . . . It's a great new feeling.

with the statement of one man,

You're thankful your health is such that you can continue working up to this point. It's a matter of concern to me right now to hang on. I'm forty-seven, and I have two children in college to support.

or with the statement of another man—a history professor,

I'm afraid I'm a bit envious of my wife. She went to work a few years ago, when our children no longer needed her attention, and a whole new world has opened to her. But myself? I just look forward to writing another volume, and then another volume.

THE WORK CAREER

Men, unlike women, perceive the onset of middle age by cues presented outside the family context, often from the deferential behavior accorded them in the work setting. One man described the first time a younger associate held open a door for him; another, being called by his official title by a newcomer in the company; another, the first time he was ceremoniously asked for advice by a younger man.

Men perceive a close relationship between life line and career line. Middle age is the time to take stock. Any disparity noted between career expectations and career achievements—that is, whether one is "one time" or "late" in reaching career goals—adds to the heightened awareness of age. One lawyer said,

> I moved at age forty-five from a large corporation to a law firm. I got out at the last possible moment, because if you haven't made it by then, you had better make it fast, or you are stuck.

There is good evidence that among men most of the upward occupational mobility that occurs is largely completed by the beginning of the middle years, or by age thirty-five (Coleman and Neugarten 1971; Jaffe 1971). Some of the more highly educated continue to move up the ladder in their forties and occasionally in their fifties. On the other hand, some men, generally the less well educated, start slipping sometime in the years from 35 to 55. The majority tend to hold on, throughout this period, to whatever rung they managed to reach. Family income does not always reflect a man's job status, for by the period of the mid-forties so many wives have taken jobs that family income often continues to rise. For this and other reasons there is considerable variation in family income in middle age as compared to earlier periods in the family cycle.

THE CHANGING BODY

The most dramatic cues for the male are often biological. The increased attention to his health, the decreased efficiency of his body, the death of friends of the same age—these are the signs that prompt many men to describe bodily changes as the most salient characteristic of middle age.

> Mentally I still feel young, but suddenly one day my son beat me at tennis.

Or,

> It was the sudden heart attack of a friend that made the difference. I realized that I could no longer count on my body as I used to do . . . the body is now unpredictable.

One forty-four-year-old added,

> Of course, I'm not as young as I used to be, and it's true, the refrain you hear so often in the provocative jokes about the decrease in sexual power in men. But it isn't so much the loss of sexual interest, I think it's more the energy factor.

A decrease in sexual vigor is frequently commented on as a normal slowing down: "my needs have grown less frequent as I've gotten older," or "sex isn't as important as it once was." The effect is often described as having little effect on the quality of the marriage.

> I think as the years go by you have less sexual desire. In fact, when you're younger there's a *need*, in addition to the desire, and that need diminishes without the personal relationship becoming strained or less close and warm. . . . I still enjoy sex, but not with the fervor of youth. . . . Not because I've lost my feelings for my wife, but because it happens to you physically.

Although there are a number of small-scale studies, the data are poor regarding the sexual behavior of middle-aged and older people. It would appear that sexual activity remains higher than the earlier stereotypes would indicate and that sexual activity in middle age tends to be consistent with the individuals' earlier behavior; but at the same time there is a gradual decrease with age in most persons, and the incidence of sexual inadequacy takes a sharp upturn in men after age fifty. . . .

Changes in health and in sexual performance are more of an age marker for men than for women. Despite the menopause and other manifestations of the climacterium, women refer much less frequently to biological changes or to concerns about health. Body monitoring is the term we used to describe the large variety of protective strategies for maintaining the middle-aged body at given levels of performance and appearance; but while these issues take the form of a new sense of physical vulnerability in men, they take the form of a rehearsal for widowhood in women. Women are more concerned over the body monitoring of their husbands than of themselves. . . .

THE CHANGING TIME PERSPECTIVE

Both sexes, although men more than women, talked of the new difference in the way time is perceived. Life is restructured in terms of time left to live rather than time since birth. Not only the reversal in directionality, but the awareness that time is finite, is a particularly conspicuous feature of middle age. Thus,

> You hear so much about deaths that seem premature. That's one of the changes that comes over you over the years. Young fellows never give it a thought.

Another said,

> Time is now a two edged sword. To some of my friends, it acts as a prod; to others, a brake. It adds a certain anxiety, but I must also say it adds a certain zest in seeing how much pleasure can still be obtained, how many good years one can still arrange, how many new activities can be undertaken.

The recognition that there is "only so much time left" was a frequent theme in the interviews. In referring to the death of a contemporary, one man said,

> There is now the realization that death is very real. Those things don't quite penetrate when you're in your twenties and you think

that life is all ahead of you. Now you know that death will come to you, too.

This last-named phenomenon we called the personalization of death: the awareness that one's own death is inevitable and that one must begin to come to terms with that actuality. . . .

THE PRIME OF LIFE

Despite the new realization of the finiteness of time, one of the most prevailing themes expressed by middle-aged respondents is that middle adulthood is the period of maximum capacity and ability to handle a highly complex environment and a highly differentiated self. Very few express a wish to be young again. As one of them said,

> There is a difference between wanting to *feel* young and wanting to *be* young. Of course, it would be pleasant to maintain the vigor and appearance of youth; but I would not trade those things for the authority or the autonomy I feel—no, nor the ease of interpersonal relationships nor the self-confidence that comes from experience.

The middle-aged individual, having learned to cope with the many contingencies of childhood, adolescence, and young adulthood, now has available to him a substantial repertoire of strategies for dealing with life. One woman put it,

> I know what will work in most situations, and what will not. I am well beyond the trial-and-error stage of youth. I now have a set of guidelines. . . . And I am practiced.

Whether or not they are correct in their assessments, most of our respondents perceive striking improvement in their exercise of judgment. For both men and women the perception of greater maturity and a better grasp of realities is one of the most reassuring aspects of being middle-aged.

> You feel you have lived long enough to have learned a few things that nobody can learn earlier. That's the reward . . . and

also the excitement. I now see things in books, in people, in music that I couldn't see when I was younger. . . . It's a form of ripening that I attribute largely to my present age.

There are a number of manifestations of this sense of competence. There is, for instance, the forty-five-year-old's sensitivity to the self as the instrument by which to reach his goals; what we have called a preoccupation with self-utilization (as contrasted to the self-consciousness of the adolescent):

I know now exactly what I can do best, and how to make the best use of my time. . . . I know how to delegate authority, but also what decisions to make myself. . . . I know how to protect myself from troublesome people . . . one well-placed telephone call will get me what I need. It takes time to learn how to cut through the red tape and how to get the organization to work for me. . . . All this is what makes the difference between me and a young man, and it's all this that gives me the advantage.

Other studies have shown that the perception of middle age as the peak period of life is shared by young, middle-aged, and older respondents (Back and Bourke 1970). In one such study there was consensus that the middle-aged are not only the wealthiest, but the most powerful; not only the most knowledgeable, but the most skillful (Cameron 1970).

There is also the heightened self-understanding that provides gratification. One perceptive woman described it in these terms:

It is as if there are two mirrors before me, each held at a partial angle. I see part of myself in my mother who is growing old, and part of her in me. In the other mirror, I see part of myself in my daughter. I have had some dramatic insights, just from looking in those mirrors. . . . It is a set of revelations that I suppose can only come when you are in the middle of three generations.

In pondering the data on these men and women, we have been impressed with the central importance of what might be called the executive processes of personality in middle age: self-awareness, selectivity, manipulation and control of the environment, mastery, competence, the wide array of cognitive strategies. We are impressed, too, with reflection as a striking characteristic of the mental life of middle-aged persons: the stocktaking, the heightened introspection, and, above all, the structuring and restructuring of experience—that is, the conscious processing of new information in the light of what one has already learned and the turning of one's proficiency to the achievement of desired ends. These people feel that they effectively manipulate their social environments on the basis of prestige and expertise; and that they create many of their own rules and norms. There is a sense of increased control over impulse life. The middle-aged person often describes himself as no longer "driven," but as the "driver"—in short, "in command."

Although the self-reports quoted here were given by highly educated and successful persons, they convey many of the same attitudes expressed less fluently by others with less education and less achievement, middle-aged people who also feel the same increasing distance from the young, the stocktaking, the changing time perspective, and the higher degrees of expertise and self-understanding.

REFERENCES

Back, K. W. and L. B. Bourque. 1970. Life graphs: aging and cohort effects. *J. Gerontol* 25:249–255.

Cameron, P. 1970. The generation gap: which generation is believed powerful versus generational members' self-appraisals of power. *Develop. Psychol.* 3:403–404.

Coleman, R. and B. L. Neugarten. 1971. *Social Status in the City*. San Francisco: Jossey-Bass.

Jaffe, A. J. 1971. The middle years. *Industrial Gerontology*, September (special issue).

Neugarten, B. L. 1967. The Awareness of middle age. In R. Owen, ed., *Middle Age*. London: British Broadcasting Co.

Twenty-Two

The Midlife Transition: A Period in Adult Psychosocial Development

Daniel J. Levinson

In an earlier selection (15), Erik Erikson discussed the identity crisis of adolescence. Is midlife a time of identity "crisis" as well? In this selection, Daniel Levinson discusses a series of psychological shifts that appear to take place during young and middle adulthood. An important turning point—or transition—appears to take place at midlife. It is this transition that might be viewed as a sort of normative identity "crisis" during adulthood.

FOR THE PAST TEN YEARS my colleagues and I have been working on a theory of adult psychosocial development in men (Levinson et al. 1974, 1977; Levinson 1977). We have started some research on women (Stewart 1977), but it is too early yet to report definitive theory or findings. Our aim is to encompass the many components of a man's life—all of his relationships with individuals, groups, and institutions that have significance for him. The components of life include his occupation and its evolution over the years, his love relationships, marriage, and family life, his various other roles and careers in numerous social contexts. This psychosocial approach includes the man's personality and the ways in which it influences and is influenced by the evolution of his careers in occupation, family, and other systems. The resulting theory is not a theory of personality development, nor of occupational development, nor of development in any single aspect of living. It deals, rather, with the development of the individual life in the broadest sense, encompassing all of these segments. This theory provides a context within which we can study in more detail the deveopment of personality and of particular careers. I shall briefly

284

describe the developmental periods we discovered in early and middle adulthood, giving major emphasis to one period, the midlife transition. Like childhood and adolescence, these periods are found in the lives of all men. Of course, men traverse them in myriad ways, as a result of differences in class, ethnicity, personality, and other factors. My primary aim is to present some of our major concepts, hypotheses, findings, and ways of thinking about adult development. None of them has been fully validated. Together, they comprise a framework for the analysis of adult development. No doubt the theory will be modified and extended as a result of further investigation.

Our primary concept is *individual life structure*. This refers to the patterning or design of the individual life at a given time. It refers to self-in-world, to the engagement of the individual in society. It requires us to consider both self and world, and the transactions between them. A life structure has three aspects:

(a) The nature of the man's *sociocultural world*, including class, religion, ethnicity, race, family, political systems, occupational structure, and particular conditions and events, such as economic depression or prosperity, war, and liberation movements of all kinds.

(b) His *participation in this world*—his evolving relationships and roles as citizen, worker, boss, lover, friend, husband, father, member of diverse groups and organizations.

(c) The *aspects of his self* that are expressed and lived out in the various components of his life; and the aspects of the self that must be inhibited or neglected within the life structure. . . .

Within a given life structure certain components are *central*: they occupy most time and energy, provide the basis on which other components are chosen, and have the greatest significance for the self. Occupation and family are usually most central in a man's life; other important components include ethnicity-religion, peer relations, and leisure. The *peripheral* components are more detachable and changeable; they involve less investment of the self and are less crucial to the fabric of one's life.

In this view, adult development is the *evolution of the life structure*. The life structure does not remain static, nor does it change capriciously. Rather; it goes through a sequence of alternating stable periods and transitional periods. The stable periods ordinarily last some six to eight years, the transitional periods four to five years.

The primary developmental task of a *stable period* is to make certain crucial choices, build a life structure around them, and seek to attain particular goals and values within this structure. Each stable period also has its own distinctive tasks, which reflect the requirements of that time in the life cycle. Many changes may occur during a stable period, but the basic life structure remains relatively intact.

The primary developmental task of a *transitional period* is to terminate the existing structure and to work toward the initiation of a new structure. This requires a man to reappraise the existing life structure, to explore various possibilities for change in the world and in the self, and to move toward the crucial choices that will form the basis for a new life structure in the ensuing stable period. Each transitional period also has its own distinctive tasks reflecting its place in the life cycle.

THE STUDY OF ADULT DEVELOPMENT

We have carried out an intensive study of forty men in the "midlife decade" (age 35–45). We started with the idea that the decade contained the shift from "youth" to "middle age." As the study progressed, we realized that we could not understand this decade in isolation, that we had to form a conception of the life cycle as a whole and of the developmental phases within it. In particular, we needed a conception of the developmental sequence from age 18 to 35 as a base for our thinking about the mid-life decade. We thus moved from the study of one decade to the study of adult development in general. Our view of eras as major components of the life cycle, and of developmental periods within and between eras, was the product of this investigation.

Fortunately, our method of research lent itself well to this expanded definition of our aims. Each man was interviewed five to ten times, for a total of ten to twenty hours. The explicitly stated task of interviewer and interviewee was to collaborate in reconstructing the story of the latter's life, with emphasis on the late adolescent and adult years.

Our method was that of *biography*: we followed the sequence of the life course, taking account of external situations, overt actions, and the ways in which they were experienced by the man and others. We tried to get at the external realities, without limiting ourselves to a purely sociological approach. Likewise, we tried to learn about the personality aspects, without limiting ourselves to a solely psycho-dynamic approach. Personality, social roles, and external circum-stances were included as aspects of the life course rather than as primary causes or explanatory variables. This is the crucial difference between a biography, on the one hand, and a clinical case history, a personality profile, or an analysis of an occupational career, on the other.

The interviews covered all segments of living that had impor-tance for the interviewee: education, work, peer relationships, family of origin, marriage and family of procreation, leisure, religion, politics. We followed each of these over the life course. Our aim was to grasp the connections among these, their patterning at a given time and the evolution of the pattern over time. This aim, vaguely defined at first, became more and more important in the conduct of the interviews and in the analysis of the life course. The biographical work generated the idea of the individual life structure, the pivotal concept in our theory of adult development.

In order to diversify the sample and to facilitate subgroup com-parisons, we selected ten men in each of four occupational categories: hourly workers in industry, business executives, academic biologists, and novelists. The men vary widely in current life circumstances, place in society, social origins, and personal characteristics. They were selected on a random basis through their work settings. The sample was minimally biased by self-selection. They were not ob-tained through a clinical facility, nor were they required to participate in the study.

I have emphasized the selection and interviewing of the sample of forty men because this sample is our primary research base. It would be a severe understatement, however, to say that this theory is based solely on a study of forty men. In addition to this primary sample, we used a larger and more diverse secondary sample drawn from many societies and historical periods.

DEVELOPMENTAL PERIODS IN EARLY ADULTHOOD

As indicated, my main emphasis here is on one of the major transi-tional periods in adult development, the midlife transition. This period ordinarily starts at age 40, give or take two years, and lasts some four to six years. In the mid-40s a new life structure emerges and persists for several years until a new transition gets underway. The midlife transition serves as a developmental link between two eras in the life cycle: early adulthood and middle adulthood. I cannot present here our overall conception of the life cycle and its component eras (see Levinson 1977; Levinson et al. 1977). Each era contains a series of developmental periods, but it also has an overall character of its own. It must suffice merely to identify the eras and their approximate age spans:

Pre-adulthood: age 0–22
Early adulthood: age 17–45
Middle adulthood: age 40–65
Late adulthood: age 60–85
Late late adulthood: age 80+

The shift from one era to the next is a massive developmental step and requires a transitional period of several years. Every transition is both an ending and a beginning, a departure and arrival, a death and rebirth, a meeting of past and future (Ortega y Gasset 1958; Unamuno 1954). The basic task of a cross-era transition is to terminate the era just ending and to create a basis for the next. Thus, the early adult transition, which lasts from roughly age 17 to 22, is part of both pre-adulthood and early adulthood. It is equally correct to say that pre-adulthood extends into the early twenties, and that early adulthood begins in the late teens. The early adult transition is of special developmental importance precisely because it contains and bridges the two eras. Likewise, the midlife transition, from roughly age 40 to 45, serves as a boundary region linking early and middle adulthood.

In order to provide a context for discussion of the midlife transition I shall first review briefly the preceding periods in early adulthood. I shall give a modal age for the onset of each period; there is a range of about two years above and below this modal figure. We did not start the research with the expectation of finding age-linked periods; they are an empirical finding rather than an *a priori* assumption.

The Early Adult Transition
This period ordinarily lasts from age 17 until 22. Its twin developmental tasks are to terminate pre-adulthood and to initiate early

adulthood. The first task is to start moving out of the pre-adult world: to question the nature of that world and one's place in it; to modify or terminate existing relationships with important persons, groups, and institutions; to reappraise and modify the self that had formed in it. Various kinds of separation, ending, and transformation must be made as one completes an entire phase of life. The second task is to make a preliminary step into the adult world: to explore its possibilities, to imagine oneself as a participant in it, to consolidate an initial adult identity, to make and test some preliminary choices for adult living.

Entering the Adult World: The First Adult Life Structure

This period extends from about age 22 to 28. The underlying developmental task is to fashion and test out a structure that provides a viable link between the valued self and the adult society. The young man must shift the center of gravity of his life from the position of child in his family of origin to the position novice adult, with a new home base that is more truly his own. He must make and live with initial choices regarding occupation, love relationships (usually including marriage and family), values, and life-style.

In this period a man has two major yet antithetical tasks: he needs to *explore* the possibilities for adult living: to keep his options open, to avoid strong commitments, and to maximize the alternatives. This task is reflected in a sense of adventure and wonderment, a wish to seek out all the treasures of the new world he is entering. The contrasting task is to *create a stable life structure:* to settle down, become responsible, and "make something of my life." Each task has sources and supports in the external world and in the self. There is usually moderate or great discontinuity between the pre-adult world in which a man grew up and the adult world in which he forms his first life structure. Exploring this new world and building a life within it is an exciting yet often confusing and painful process.

Each man works out his own balance of exploration-openness and stability-commitment during this period. Some men go through it on a highly tentative basis, leading transient lives and capriciously changing jobs, residences, relationships. They create a loose structure without stability or rootedness—not investing much of the self in the world nor taking much of the world into the self. Others have some mixture of stability and change, making some commitments but feeling open to change. Or they initially "keep loose" and then at perhaps age 25 or 26 get more serious about forming a stable life. A man may make a strong commitment in one sector of his life, such as work or marriage, and not in others.

Still other men make strong commitments at the outset and build what they hope will be an enduring life structure. These men usually make their key choices regarding marriage and occupation in the early adult transition and maintain great continuity with the pre-adult world, without exploring new options or questioning their values and goals.

In the late 20s, as this period draws to a close, the seemingly firm structures comes into question. The man wonders: Did I commit myself prematurely? Do I want to keep this way of life forever? Shall I seek new possibilities more suitable for the person I am now? Like-wise, the young man who had created a loose or fragmented structure now comes to recognize its limitations and to seek a life of greater order and more stable attachments.

The Age Thirty Transition

This transition, which lasts from about age 28 to 33, is another of nature's ingenious devices for permitting growth and redirection in our lives. It provides an opportunity to work on the flaws and lim-itations of the first adult life structure and to create the basis for a new and more satisfactory structure with which to complete the era of early adulthood.

At about 28 the provisional, exploratory quality of the 20s is ending and life is becoming more serious, more restrictive, more "for real." A voice from deep within the self says: "If I want to change my life—if there are things in it I want to modify or exclude, or things missing I want to add—I must now make a start, for soon it will be too late." Men differ in the kinds of changes they make, but the life structure is always different at the end of the age thirty transition than it was at the start.

Some men have a rather smooth transition, without overt dis-ruption or sense of crisis. They may use this period to modify and enrich their lives in certain respects, but they build directly upon the past and do not make fundamental changes. It is a time of reform, not revolution.

For other men the age thrity transition takes a more stressful form, the *age 30 crisis*. A truly developmental crisis occurs when a man is having difficulty with the developmental tasks of this period;

he finds his present life structure is intolerable, yet seems unable to form a better one. In a severe crisis he experiences a threat to life itself, the danger of chaos and dissolution, the loss of hope for the future. Most of the men in our study went through a moderate or severe crisis during this period.

There is a peaking of marital problems and divorce in the age thirty transition. It is also a time for various kinds of occupatinal change: a shift in occupational category or kind of work; a settling down after a period of transient or multiple jobs; a promotion or advancement that brings one into a new occupational world; and, apart from the external aspects, a change in the internal meanings and goals of work. Many men hit "rock bottom" in this period before discovering a new path. There is a higher frequency of psychotherapy. For some men the age thrity transition is a period of excitement, of growth, of "getting oneself together" for the first time.

These first three periods—the early adult transition, entering the adult world, and the age thrity transition—generally last about fifteen years, from age 17–18 until 32–33. Together, they constitute the preparatory, "novice" phase of early adulthood. This is not an extended adolescence but a highly formative, evolving phase of adult life (White 1952).

The shift from the end of the age thirty transition to the start of the next period is one of the crucial steps in adult development. At this time a man may make important new choices, or he may reaffirm old choices. If these choices are congruent with his dreams, values, talents, and possibilities, they provide the basis for a relatively satisfactory life structure. If the choices are poorly made and the new structure seriously flawed, however, he will pay a heavy price in the next period. But even the best structure has its contradictions and must in time be changed.

Settling Down: The Second Adult Life Structure

The second life structure takes shape at age 32 or 33 and persists until age 39 or 40. A young man now tries to build a life structure for the culmination of early adulthood. He seeks to invest himself in the primary components of this structure (work, family, friendships, leisure, community—whatever is most central for him) and to realize his youthful aspirations.

In this period a man has two major tasks: to *establish a niche* in society: to dig in, anchor his life more firmly, develop competence in a chosen craft, become a valued member of a valued world; to work at *"making it"*: planning, striving to advance, progressing along a timetable. One version of "making it" is the American success dream of becoming rich, powerful, and famous. I use the term more broadly to include all efforts to build a better life, to attain valued goals, to contribute to the tribe and be affirmed by it. The artist and intellectual, like the worker, businessman, cleric, and revolutionary, normally use the settling down period toward these aims.

Until the early 30s, a young man has been a "novice" adult, forming an adult life and seeking a more established place in adult society. His task in the settling down period is to become a full-fledged adult within his own world. He defines a *personal enterprise*, a direction in which to strive, a sense of the future, a life plan, a "project" as Sartre (1953) has termed it. The enterprise may be precisely defined from the start (to become a successful middle manager or corporation executive, to become a great scientist or artist, to have some financial security and a happy family life in the suburbs) or it may take shape only gradually over the course of this period.

The imagery of the *ladder* is central to the settling down enterprise. The ladder is psychosocial: it reflects the realities of the external social world, but it is defined by the person in terms of his own meanings, motives, and strivings. By "ladder" I mean all dimensions of advancement—social rank, income, power, fame, creativity, quality of family life, social contribution—as these are important for the man and his world.

At the start of this period, a man is on the bottom rung of his ladder and is entering a world in which he is a *junior member*. His aims are to advance in the enterprise, to climb the ladder and become a *senior member* in that world. His sense of well-being during this period depends strongly on his own and others' evaluation of his progress toward these aims.

Toward the end of the settling down period, starting ordinarily at age 36–37, there is a phase so distinctive that we have given it a name: *Becoming One's Own Man*. The acronym BOOM connotes an important aspect of what man wants in this phase, though not necessarily what he gets. The major developmental tasks of BOOM are to accomplish the goals of the settling down enterprise, to become a senior member in that world, to speak more strongly with one's own voice, and to have a greater measure of authority. A crucial dilemma here is that a man wants to be more independent, more true to his own wishes even if they run counter to external demands, but he also wants affirmation, respect, and reward from his world.

This is a fateful time in a man's life. Attaining seniority and approaching the top rung of his ladder are signs to him that he is becoming more fully a man (not just a person, but a male adult). However, his progress not only brings new rewards, it also carries the burden of greater responsibilities and pressures. It means that he must give up even more of the little boy within himself—an internal figure who is never completely outgrown, and certainly not in early adulthood—and this intensifies the inner struggle between the little boy and the evolving man.

The activation of the pre-adult self during BOOM is a part of normal psychosocial development. It represents a step forward in that it brings out the boy-man polarity and creates the possibility of re-solving it at a higher level. For a time, however, the conflicts may contribute to difficulties in the man's relationships with wife, children, lover, boss, friends, coworkers. He often feels that others regard him as a little boy to be deprived, exploited, and controlled. He may experience his boss as a tyrannical, corrupt, or inept authority, his wife as a depriving or smothering mother, his critics as jealous rivals who cannot tolerate his success. There is usually some mixture of reality and distortion in these perceptions; it takes time to deal with the inner conflicts and to gain a more realistic view of self and others. . . .

In some cases, settling down begins with a seemingly stable structure, but as BOOM gets under way, the man experiences more acutely the flaws in this structure and his life becomes intolerable. The locus of difficulty may be in one sector such as work or marriage, or he may feel profoundly alienated from his entire world—suffocated, oppressed, and without space in which to be himself and to do what matters most. He decides in BOOM that he must change his life. Our name for this sequence is *Breaking Out.* He seeks a radical change in life structure. No matter how well his enterprise succeeds, the sucess will not provide what he really wants: He is in the wrong enterprise, involved in the wrong relationships, violating his most cherished dreams and aspirations. He decides to build a new life within which he can truly become his own man and live out the most important aspects of his self.

The breaking out may be dramatized by a decisive act such

as quitting his job, leaving his wife, or moving to "another part of the country." However, this event is but one step in a developmental process that began earlier and will continue for some years. It is part of a "BOOM crisis" that is as fundamental as the other, more documented crises in human development. This crisis involves a terrible dilemma. The man fears that if he stays put, he faces a living death, a future without promise, a self-betrayal. Yet to make a major change at this point may be hurtful to his loved ones as well as himself; his hopes for a better life may be only a pipe dream that will soon end in failure, disappointment, and remorse.

Breaking out, does, indeed, exact great costs. A man who attempts a major change at age 36 or 38 is not likely to establish a new structure and gain some stability in his life in less than eight to ten years. It takes time to break away, to sever or modify old relationships and commitments. It takes time, also, to "break in," to enter a new world and create a new life. The new enterprise is hardly under way when the midlife transition begins and, with it, a new effort to question and improve one's life.

For every man who attempts to break out, there are several who, after great inner struggle, decide to stay put. The long-term costs of remaining in an oppressive life structure may also be very large. It is a task for Solomon—to say, in any individual case, what the wise decision is.

THE MIDLIFE TRANSITION

During his 30s, a man's primary developmental task is to build a satisfactory life structure. Within it, he tries to establish his place in society, to pursue his youthful aspirations, and to live out important aspects of the self. The late 30s mark the culmination of early adulthood. The midlife transition, which lasts from roughly age 40 to 45, provides a bridge from early to middle adulthood.

The start of the midlife transition brings a new set of developmental tasks. Now the life structure itself comes into question and cannot be taken for granted. It becomes important to ask: What have I done with my life? What do I really get from and give to my wife, children, friends, work, community—and self? What is it I truly want for myself and others? What are my real values and how are they reflected in my life? What are my greatest talents and how am I using—or wasting—them? What have I done with my early dreams and what do I want with them now? Can I live in a way that best combines my current desires, values, talents, and aspirations?

Some men—but very few, according to our study—do very little questioning or searching during the midlife transition. They are apparently untroubled by difficult questions regarding the meaning, value, and direction of their lives. They may be working on these questions in implicit or unconscious ways that will become evident later. If not, they will pay the price in a later developmental crisis or a life structure minimally connected to the self. Other men, again not a large number, realize that the character of their lives is changing, but the process is not a painful one. They are, so to say, in a manageable transition without crisis.

For the great majority—about 80 percent of our subjects—this period evokes tumultuous struggles within the self and with the external world. Their midlife transition is a time of moderate or severe crisis. They question virtually every aspect of their lives and are horrified by what they find. They cannot go on as before, but it will take several years to form a new path or to modify the old one.

Because a man in this crisis is often somewhat irrational, others may regard him as "upset" or "sick." It is therefore important to recognize that he is working on a normal midlife task and that the desire to question and enrich his life stems from the most healthy part of the self. The doubting and searching are "normal"; the real question is how best to make use of them. The difficulty is compounded by the fact that the process of reappraisal activates neurotic problems (the unconscious baggage carried forward from hard times in the past) which hinder the effort to change. The pathology is not in the desire to improve one's life but in the inner obstacles to pursuing this aim. The pathological anxiety and guilt, the ancient dependencies, animosities, and vanities of earlier years, keep one from examining the real issues at midlife. They make it difficult for a man to free himself from an oppressive life structure.

A profound reappraisal of this kind cannot be a cool, intellectual process. It must also involve emotional turmoil, despair, the sense of not knowing where to turn or of being stagnant and unable to move at all. A man in this state often makes false starts. He tentatively tests a variety of new choices, not only out of confusion or impulsiveness but also out of a need to explore, to learn how it feels to engage in a particular love relationship, occupation, or solitary pursuit. Every genuine reappraisal must be agonizing because it challenges the assumptions, illusions, and vested interests on which the existing structure is based.

The life structure of the 30s was initiated and stabilized by powerful forces within the person and in his environment. These forces continue to make their claim for preserving the status quo. The man who attempts a radical critique of his life at age 40 will be up against the parts of himself that have a strong investment in the present structure. He will often be opposed by other persons and institutions—his wife, children, boss, parents, colleagues, the organization and the broader occupational system in which he works, the implicit web of social conformity—that seek to maintain order and prevent change. With luck, he will also find support in himself and others for the effort to examine and improve his life.

Why do we go through this painful process? We need developmental transitions in adulthood partly because no life structure can permit the living out of all aspects of the self. To create a life structure I must make choices and set priorities. In making a choice, I select one option and reject many others. Committing myself to a structure, I try over a span of time to realize its potential, to bear the responsibilities and tolerate the costs it entails.

The life structure of the 30s necessarily gives high priority to certain aspects of the self and neglects or minimizes other aspects. With the onset of midlife transition, these neglected parts of the self urgently seek expression and stimulate a man to reappraise his life. He experiences them as "other voices in other rooms" (to build on Truman Capote's evocative phrase), as internal voices that have been silent or muted for years and now clamor to be heard. He may hear only a vague whispering, the content unclear but the tone indicating grief over lost opportunities, outrage over betrayal by others, or guilt over betrayal by himself. Or the voices may come through as a thunderous roar, the content all too clear, stating names and times and places and demanding that something be done to right the balance. He hears the voice of an identity prematurely rejected; of a love lost or not pursued; of a valued interest or relationship given up in acquiescence to parental or other authority; of an internal figure who wants to be an athlete or nomad or artist, to marry for love or remain a bachelor, to get rich or enter the clergy or live a sensual carefree life—possibilities set aside earlier to become what he now is. During the midlife transition, he must learn to listen more attentively to these voices and, in the end, decide what part he will give them in his life.

In the effort to reappraise his life, the man discovers how much it has been based on illusions, and he is faced with the task of *de-illusionment*. By this I mean the reduction of illusions, the recognition that long-held assumptions and beliefs about self and world are not true. . . . The process of de-illusionment involves many feelings such as disappointment, joy, relief, bitterness, grief, wonder, freedom. It has diverse outcomes: the person may become disillusioned in the sense of suffering an irreparable loss; he may also become free to form more flexible values and to admire others in a more genuine, less idealizing way. . . .

As the midlife transition nears its end in the middle 40s, a man has to make new choices or recommit himself on different terms to old choices. If he is to improve his original marriage, or enter a new one that will be an improvement on the old, he must become less illusioned about himself. He has to accept some responsibility for those aspects of his own motivation and character that keep him from having more adult relationships with women. He has to work over a period of several years to develop a new kind of relationship with a woman who is ready to join him in this mutual effort. Otherwise he will remain in a stagnant marriage destructive to both partners, or he will embark upon a new marriage (or a new set of relationships with women) that repeats the old hurtful themes with new variations.

If he is to make significant changes in love, occupation, leisure, and other important aspects of his life, a man must engage in the process of midlife individuation. In particular, he must confront and integrate further the great polarities that need to be reworked in every developmental transition: young and old, masculine and feminine, destructiveness and creativeness, attachment and separateness. These archetypal polarities represent basic divisions in the individual and in society. They can never be entirely integrated or transcended. In successive developmental periods from infancy through old age, one has an opportunity to reduce the internal splitting and to find new ways of being young/old, masculine/feminine, destructive/creative, and attached/separate, according to one's place in the life cycle.

The life structure that emerges in the middle 40s varies greatly in its satisfactoriness, that is, in its suitability for the self and its viability in the world. Some men have suffered such irreparable defeats in childhood or early adulthood, and have been able to work so little on the tasks of their midlife transition that they lack the inner and outer resources for creating a minimally adequate structure. They face a middle adulthood of constriction and decline. Other men form a life structure that is reasonably viable in their world. They keep busy, perform their social roles, and do their bit for themselves and others. However, this structure is poorly connected to the self, and their lives are lacking in inner excitement.

Still other men have made a start toward a middle adulthood that will have its own special satisfactions and fulfillments as well as burdens. For these men, middle adulthood is often the fullest and most creative season in the life cycle. They are less tyrannized by the ambitions, instinctual drives, and illusions of youth. They can be more deeply attached to others and yet more separate, more centered in the self. For them, the season passes in its proper rhythm.

The midlife transition is not the last opportunity for change and growth. Work on our developmental tasks can continue through middle adulthood and beyond, and there are later transitional periods to facilitate the process. As long as life continues, no period marks the end of the opportunities and the burdens of further development.

REFERENCES

Levinson, D. J. 1977. The changing character of middle adulthood in American society. In G. J. DiRenzo, ed., *We the People: Social Change and Social Character* Westport, Conn.: Greenwood Press.

Levinson, D. J. et al., 1974. The psychosocial development of men in early adulthood and the midlife transition. In D. F. Ricks et al., eds., *Life History Research in Psychopathology*. Minneapolis: University of Minnesota Press.

Levinson, D. J. (in collaboration with C. M. Darrow, E. B. Klein, M. H. Levinson, and B. McKee). 1977. *Seasons of a Man's Life*. New York: Knopf.

Ortega Y Gasset, J. 1958. *Man and Crisis*. New York: Norton.

Sartre, J. P. 1953 *Existential Psychoanalysis*. New York: Philosophical Library.

Stewart, W. A. 1977. The formation of the early adult life structure in women. Ph.D. dissertation, Columbia University.

Unamuno, M. de. 1954. *Tragic Sense of Life*. New York: Dover.

White, R. 1952. *Lives in Progress*. New York: Dryden.

Adulthood
Relations with Others

Twenty-Three

Marital Satisfaction over the Family Life Cycle

Boyd C. Rollins and Harold Feldman

In this selection, Rollins and Feldman discuss changes in marital satisfaction over the course of the family life cycle. Their findings suggest that marital satisfaction is influenced to a great extent by events outside the marital relationship. For women, marital satisfaction appears to change primarily as a function of shifts in parenting responsibilities; for men, marital satisfaction may be influenced more by events in the occupational sphere.

THE CONCEPT of developmental adjustment in marriage has recently stimulated concern for patterns of change in marital interaction over the family life cycle. . . . The issue of the meaning of marital success, its pattern over the family life cycle and its developmental antecedents and correlates is a primary issue in contemporary family sociology. . . .

According to Burgess and Locke (1945:439), "satisfaction appears to be a correspondence between the actual and the expected or a comparison of the actual relationship with the alternative, if the present relationship were terminated." Such a definition permits a focus on the total marital relationship or on specific aspects, on discrepancies between role expectations and perceived role performances, or between goals and perceived goal attainment, or between the personal qualities in one's conception of the ideal spouse and his perception of the actual personal qualities of his spouse. The meaning of such criteria is in the subjective evaluation of the individual participant from his particular point of view. Therefore, it is possible that psychological events such as perception of a values discrepancy with spouse when there is actually none, as well as the realities of married life circumstances over the family life cycle, influence the pattern of marital satisfaction or dissatisfaction. Also, the likelihood of such psychological events as well as their relationship to marital satisfaction might be very different for men than for women.

As a descriptive device the family life cycle has been used to compare structures and functions of marital interaction in different stages of development. A fairly simple scheme and the one used in this study was to classify couples into eight stages of the family life cycle in terms of the age of the oldest child similar to what Duvall (1967:9) had done. The classification was as follows:

Stage I. Beginning families (couples married 0 to 5 yrs. without children)

Stage II. Childbearing families (oldest child, birth to 2 yrs. 11 mos.)

Stage III. Families with preschool children (oldest child, 3 yrs. to 5 yrs. 11 mos.)

Stage IV. Families with school-age children (oldest child, 6 yrs. to 12 yrs. 11 mos.)

Stage V. Families with teenagers (oldest child, 13 yrs. to 20 yrs. 11 mos.)

Stage VI. Families as launching centers (first child gone to last child's leaving home)

Stage VII. Families in the middle years (empty nest to retirement)

Stage VIII. Aging families (retirement to death of first spouse). . . .

It was hoped that a substantiated pattern of marital satisfaction would be described from this study and provide a basis for beginning attempts to construct a developmental theory of marital satisfaction.

METHOD

Data for this study were obtained through the use of an area survey sample of middle class residents of Syracuse, New York, in 1960. Dr. Charles Willie, a sociologist at the University of Syracuse had previously classified all the census tracts in the city of Syracuse, New

York, into one of six "social areas" in terms of percent of single family dwellings, average monthly rental, average market value of owned homes, median number of school years completed, and percent of operatives, service workers, and laborers in the census tract. The census tracts in the top two social areas were considered to include a large proportion of upper-middle-class and upper-class residents.

The nine census tracts in the top two socioeconomic categories of the city were sampled in this study.

Field workers stopped at each selected housing unit, left a questionnaire for each husband and each wife, and made an appointment to pick them up within a few days. The questionnaire asked for information on family of orientation and family of procreation, marital history, occupation, marital satisfaction, communication, decision-making, methods of handling conflicts, values, frequency of integrative and disruptive experiences, and satisfaction with stages of the family life cycle from each individual. Only the data used to classify couples by stage of family life cycle and marital satisfaction are analyzed in this paper. A high response rate of 85 percent usable questionnaires from both husbands and wives in the same households were collected from the target housing units. This provided data on a total of 852 married couples.

On religious preference 21 percent of the couples indicated Catholic, 35 percent Protestant, 27 percent Jewish, and 17 percent either mixed or none. Eighty-eight percent of the husbands were classified as white collar and only 12 percent as blue collar according to their occupation, and only 12 percent of the couples included a person who had been married previously to another person. Sixty-eight percent of the husbands had received some college education and 24 percent had received postgraduate education. The sample was predominately Caucasian, well-educated, middle- and upper-class persons in their first marriage with the wife not working outside the home.

Fifty-three of the initial 852 couples were married for more than five years and were still childless. They were eliminated from the analysis because they were considered atypical in terms of stages of the family life cycle. On the basis of length of time married, age of oldest child, and residence of children, the remaining 799 couples were classified into one of eight stages of the family life cycle. The distribution of these couples was as follows: 51 at Stage I ("Beginning"), 51 at Stage II ("Infant"), 82 at Stage III ("Preschool"), 244 at

Stage IV ("School-age"), 227 at Stage V ("Teenage"), 64 at Stage VI ("Launching"), 30 at Stage VII ("Empty nest"), and 50 at Stage VIII ("Retirement").

The data on marital satisfaction were taken from four questions on the questionnaire as follows:

1. General Marital Satisfaction—"In general, how often do you think that things between you and your wife are going well? —all the time, —most of the time, —more often than not, —occasionally, —rarely, —never."

2. Negative Feelings from Interaction with Spouse—"How often would you say that the following events occur between you and your husband (wife)? —never, —once or twice a year, —once or twice a month, —once or twice a week, —about once a day, —more than once a day." The combined responses of each individual in reply to "you feel resentful," "you feel not needed," and "you feel misunderstood" were used in the data analysis. In a factor analysis of the data these three events were equally highly loaded on a dominant factor.

3. Positive Companionship Experiences with Spouse—"How often would you say that the following events occur between you and your husband (wife)? —never, —once or twice a year, —once or twice a month, —once or twice a week, —about once a day, —more than once a day." The combined responses of each individual in reply to "laugh together," "calmly discuss something together," "have a stimulating exchange of ideas," and "work together on a project" were used in the data analysis. In a factor analysis of the data these four events were equally and highly loaded on a dominant factor.

4. Satisfaction with Present Stage of the Family Life Cycle—"Different stages of the family life cycle may be viewed as being more satisfying than others. How satisfying do you think the following stages are? —very satisfying, —quite satisfying, —somewhat satisfying, —not satisfying. Data on this question were used in reply to "before the children arrive" for individuals in Stage I, "first year with infant" for Stage II, "preschool children at home" for Stage III, "all children at school" for Stage IV, "having teenagers" for Stage

V, "children gone from home" for Stages VI and VII, and "being grandparents" for Stage VIII.

The data analysis consisted of determining cross tabulation frequencies on the response categories for each of the four questions on marital satisfaction by stages of the family life cycle. This was done separately for each sex. Chi square was computed from the cross tabulation frequencies to test the null hypothesis that each stage of the family life cycle would have the same proportion of response frequencies for each of the response categories on marital satisfaction. . . .

FINDINGS

The distribution of scores on "general marital satisfaction" was similar to that of many studies (cf. Hamilton 1929, Bernard 1934; Burgess and Cottrell 1939; Burgess and Wallin 1953; Lang 1953; Bossard and Boll 1955) using either a single item measurement or a composite index. The majority of subjects were on the high end of the scale. Eighty percent of the wives and 80 percent of the husbands indicated that things were going well in their marriage all of the time or most of the time (see tables 23.1 and 23.2).

The pattern of "general marital satisfaction" for wives was a steady decline from the "beginning" to the "school-age" stage; then a leveling off with a rapid increase from the "empty nest" to the "retired" stage of the family life cycle (see figure 23.1). For the husbands there was a slight decline from the "beginning" to the "school-age" stage, a slight increase to the "empty nest" stage and then a rapid increase to the "retired" stage. The amount of change for husbands, though statistically significant, was much less than that for wives. However, this is accounted for in part by the fact that a greater percent of wives than husbands report their marriage to be "perfect" in Stage I of the family life cycle. The greatest fluctuation in marital satisfaction for husbands was the increase from Stage VII to Stage VIII. As indicated in tables 23.1 and 23.2, the chi square and the contingency coefficient values were higher for wives than husbands.[1] . . .

[1] In the reanalysis of these data, it was found that only 4 percent of the variation in marital satisfaction could be associated with variation in the family life cycle.

Table 23.1. Percentage Distribution of Wives by Stage of the Family Life Cycle and by Level of Marital Satisfaction for Four Measures of Marital Satisfaction

Measure and Level of Marital Satisfaction	Stage of Family Life Cycle								Total	Statistical Evaluation
	I N = 51	II N = 51	III N = 82	IV N = 244	V N = 227	VI N = 64	VII N = 30	VIII N = 50	N = 799	
General Marital Satisfaction										
All the time	41%	31%	22%	11%	14%	20%	17%	38%	20%	$x^2 = 55.8$
Most of the time	47	51	58	63	55	56	43	50	56	df = 14
Less often*	12	18	20	26	31	24	40	12	24	$p < .001$
										C = .31
Negative feelings										
Never	10%	4%	4%	8%	12%	25%	13%	28%	11%	$x^2 = 61.9$
Once-twice a year	41	37	40	36	49	42	54	44	42	df = 21
Once-twice a month	35	41	45	39	26	20	30	18	33	$p < .001$
More often*	14	18	11	17	13	13	3	10	14	C = .31
Positive Companionship										
More than once a day	16%	10%	7%	5%	5%	5%	10%	12%	7%	$x^2 = 46.0$
About once a day	55	39	29	31	36	38	27	24	35	df = 21
Once-twice a week	25	39	49	46	34	45	40	44	50	$p < .001$
Less often*	4	12	25	18	25	12	23	20	18	C = .27
Present FLC Stage										
Very satisfying	74%	76%	50%	35%	17%	8%	17%	82%	45%	$x^2 = 242.2$
Quite satisfying	22	18	33	44	38	16	13	14	33	df = 14
Less satisfying*	4	6	17	21	15	76	70	4	22	$p < .001$
										C = .59

* Two or more categories were combined for statistical analysis.

Table 23.2. Percentage Distribution of Husbands by Stage of the Family Life Cycle and by Level of Marital Satisfaction for Four Measures of Marital Satisfaction

Measure and Level of Marital Satisfaction	Stage of Family Life Cycle								Total	Statistical Evaluation
	I $N = 51$	II $N = 51$	III $N = 82$	IV $N = 244$	V $N = 227$	VI $N = 64$	VII $N = 30$	VIII $N = 50$	$N = 799$	
General Marital Satisfaction										
All the time	27%	22%	17%	14%	18%	27%	27%	42%	20%	$x^2 = 32.5$
Most of the time	61	62	59	63	60	55	40	52	60	df = 14
Less often*	12	16	24	23	22	18	33	6	20	$p < .01.$ $C = .24$
Negative Feelings										
Never	10%	10%	7%	15%	19%	12%	10%	32%	15%	$x^2 = 32.4$
Once-twice a year	41	47	54	50	44	63	43	40	48	df = 21
Once-twice a month	31	31	28	25	28	17	37	22	27	$p < .05$
More often*	18	12	11	10	9	8	10	6	10	$C = .23$
Positive Companionship										
More than once a day	22%	8%	4%	6%	8%	10%	10%	6%	8%	$x^2 = 42.2$
About once a day	49	43	34	35	34	31	40	26	36	df = 21
Once-twice a week	27	37	38	41	39	45	33	60	40	$p < .01$ $C = .26$
Less often*	2	12	24	18	19	14	17	8	16	$x^2 = 184.7$
Present FLC Stage										
Very satisfying	55%	69%	61%	39%	44%	9%	24%	66%	44%	df = 14
Quite satisfying	39	23	31	45	41	25	13	30	37	$p < .001$
Less satisfying*	6	8	8	16	15	66	63	4	19	$C = .53$

*Two or more categories were combined for statistical analysis.

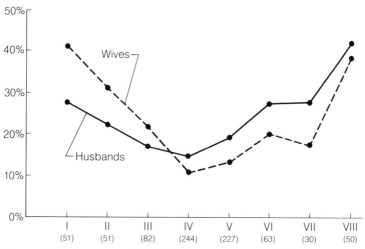

Figure 23.1. Percentage of individuals at each stage of the family life cycle (from Stage I, "beginning families," to Stage VIII, "retirement") reporting their marriage was going well "all the time." (Figures in parentheses indicate the number of husbands and also the number of wives in each stage. There was a total of 1598 cases.)

SUMMARY AND CONCLUSIONS

On only two of the four indices of marital satisfaction is there a consistent pattern over the family life cycle for both husbands and wives. Concerning the frequency of positive companionship experience, they both reported a substantial decline from the beginning of marriage to the "preschool" stage and then a leveling off over the remainder of the stages. The events used to form this scale, laughing together, calm discussions with each other, having a stimulating exchange of ideas with each other, and working together on a project, are events to which they could both objectively report a frequency of occurrence. Since they were mutual events, we would expect a similar response from the husbands and wives in each couple, if they were objective in their evaluations. The pattern here seems to be very clear—that stimulating common activity in marriage decreases from the very beginning with no recovery. . . .

Concerning satisfaction with present stage of the family life cycle, both husband and wife rate highly the childbearing and early child-rearing phases and are at a low point when launching the children from the home. Perhaps this is an indication of satisfaction with parenthood more than marriage.

The two indicies of marital satisfaction in which husbands and wives follow different patterns over the family life cycle refer to the subjective affective state of each individual with reference to their marriage. In general, husbands seem to be much less affected by stage of the family life cycle in their subjective evaluations of marital satisfaction than are wives. The husbands vary little from the establishment through the childbearing and child-rearing phases. However, the wives have a substantial decrease in general marital satisfaction and a high level of negative feelings from marital interaction during the childbearing and child-rearing phases until the children are getting ready to leave home. After the child-rearing phases both husbands and wives have a substantial increase in marital satisfaction through the "retirement" stage with an apparent temporary setback just before the husband retires.

These data suggest that experiences of childbearing and child-rearing have a rather profound and negative effect on marital satisfaction for wives, even in their basic feelings of self-worth in relation to their marriage. Perhaps this is partly a consequence of the great reduction in positive companionship experiences with their husbands instigated by the pressures of child-rearing responsibilities. On the other hand, the loss of companionship seems to occur for husbands without a decrease in marital satisfaction. The most devastating period of marriage for males appears to be when they are anticipating retirement. Marital satisfaction might be influenced more by occupational experiences for husbands than the event and developmental level of children in their families.

These data suggest that marriage has very different meanings for husbands than for wives and that very different events within or outside the marriage and/or family influence the developmental pattern of marital satisfaction in men and women. This might help explain the fact that some studies have found family life cycle differences for both men and women and some for women only. It seems that men are influenced more by events both before and after the arrival of children, while women are influenced more by the presence of children.

From a review of the literature and the data reported in this study, it seems evident that marital satisfaction of husbands and wives is associated with stages of the family life cycle, and a developmental theory of marital satisfaction is needed to explain this association. However, it is questionable that the same developmental theory would have utility for both husbands and wives. It is suggested that a developmental theory of marital satisfaction for wives would focus on the contingent role of parenthood, while for husbands the contingent occupational role seems more relevant.

REFERENCES

Bernard, Jessie. 1934. Factors in the distribution of success in marriage. *American Journal of Sociology* 40:49–60.

Bossard, James H. S. and Eleanore S. Boll. 1955. Marital unhappiness in the life cycle of marriage. *Marriage and Family Living* 17:10–14.

Burgess, Ernest W. and Leonard S. Cottrell, Jr. 1939. *Predicting Success or Failure in Marriage*. New York: Prentice-Hall.

Burgess, Ernest W. and Harvey J. Locke. 1945. *The Family: From Institution to Companionship*. New York: American Book Company.

Burgess, Ernest W. and Paul Wallin. 1953. *Engagement and Marriage*. Philadelphia: Lippincott.

Duvall, Evelyn M. 1967. *Family Development*. Philadelphia: Lippincott.

Hamilton, Gilbert V. 1929. *A Research in Marriage*. New York: Boni.

Lang, Robert O. 1953. The rating of happiness in marriage. In Ernest W. Burgess and Willard Waller, eds., *Engagement and Marriage*, pp. 536–539. Philadelphia: Lippincott.

Twenty-Four

Primary Friends and Kin: A Study of the Associations of Middle-Class Couples

Nicholas Babchuk

Little is known about the nature and function of friendship during adulthood. In this study, Nicholas Babchuk examines influences on the friendship patterns of middle-class couples. His findings indicate that friendship among middle-class adults is very couple-oriented, that husbands appear to influence friendship choices more than wives, and that, in general, adults have fewer friends in their local community than might be expected.

THE PRESENT REPORT is concerned with the primary friendships of middle-class couples. Previously it was found that husbands exercised greater influence in determining who the primary friends of the couple would be. Also, friendships initiated by the husband were more primary than those (lesser in number) initiated by the wife (Babchuk and Bates 1963). Information from two additional samples (one purposive, the other random) having somewhat different characteristics from the original population has now been gathered to provide more detailed information bearing on the initial findings and to break new ground and make possible a more refined analysis of primary relations. Thus, as in the earlier research, spouses were asked to identify mutual primary friends, indicate who had initiated such friendships, note the types of activities engaged in, provide information on the frequency of visiting, and so forth. But, in a different vein, information was sought on why couples disagreed in the friendship choices they made, on the frequency with which couples saw relatives (assuming that this fact would bear on the frequency with which couples would see friends), and on the role played by children in the

311

friendship patterning of the couple. Finally, the question was raised as to whether spouses maintained primary friends independent of each other.

A "primary friend" was a person with whom one was predisposed to enter into a wide range of activities (within limits imposed by such factors as interests, sex, age, financial resources, etc.), and with whom there would be a predominance of positive affect (Bates and Babchuk 1961). . . .

It was assumed that couples in the two new samples would provide further confirmation for the hypothesis that "the husband will initiate a greater number of primary friendships shared by the couple than will his wife. (Very close friends of the male in the period prior to marriage are more likely to become mutual friends of the pair after marriage than very close friends of the female. Also mutual friends developed subsequent to marriage are more likely to be introduced to the pair by the husband rather than wife.)" It was expected that interaction with primary friends by the couple would be a function of the amount of participation engaged in with relatives. Couples who had extensive and frequent contact with kin, it was reasoned, would be less likely to maintain extensive and frequent contact with primary friends because they would have less need for such association. Similarly, spouses who had close friends apart form those shared might, as a result, have fewer friends in common.

SAMPLE AND METHOD

The original 39 couples studied (Sample A) were white, college educated, relatively young (20 to 40 years of age), and had been married for a short time (24 couples had been married less than three years). Most of them resided in Lincoln, Nebraska. Two samples were selected for the present study. The first of these (Sample B) included 39 couples that were comparable to Sample A in educational achievement, race, and place of residence. However, a deliberate effort was made to include persons who were older and who had been married for a longer period of time. The age range in Sample B was 19 to 58. Sixteen couples had been married less than three years, 8 had been married from three to less than ten years, 8 from ten years to less to than twenty, and the remaining 7 had been married from twenty to thirty-six years. Finally, 39 couples were randomly selected (Sample C) from a middle-class, homogeneous census tract in Lincoln. All households in this tract were identified and a table of random numbers was used in drawing the sample. Persons in Sample C were 22 to 55 years of age. Only 3 couples had been married less than three years, another 14 had been married less than ten years, 16 had been married from ten to twenty years, and 6 couples had been married from twenty to twenty-seven years. This sample compared favorably in educational achievement and occupation to Samples A and B.

A structured schedule was used. For the most part the same information was sought from all three samples; however, a greater amount of information was requested from individuals in Samples B and C. The time required to interview a couple varied from one and one-half to four hours depending on the number of friends cited. The average interview lasted two and one-half hours. Respondents provided information freely. Most subjects became obviously interested in the area under study once the interview had begun; many became especially involved (and sometimes anxious) in that part of the interview where the lists of close mutual friends of husband and wife were compared and discussed. . . .

THE FINDINGS

In all three samples, husbands were dominant in determining mutual friendships whether the couples were in complete, high, or low agreement. . . .

A detailed breakdown of the friendships of the 117 couples revealed 867 friendship units cited by either husband or wife. Both spouses listed 367 friendship units in common, husbands listed 246 units not listed by their wives, and wives listed 254 units not cited by their husbands. Of the mutual friendships listed in common husbands initiated friendships in 221 instances, wives in 82 cases, and neither spouse could be credited with initiating the relationship in 54 instances. This latter category ordinarily included persons whom the couple met at the same time and who eventually became primary friends. Among the 246 units cited only by the husbands, they claimed credit for initiating the friendships in 1972 cases, credited their wives with establishing the friendships in 46 cases, and claimed neither was instrumental in the remaining 28 cases. Although men were dominant in establishing primary friendships, they may have had a somewhat exaggerated view of their influence. In this regard, for example, of 254 friendships cited only by wives, the wives claimed credit for initiating the friendships in 114 cases, credited their husbands with initiating 88 friendships, and indicated that neither spouse was instrumental in 52 cases. Friendships cited only by wives considered in conjunction with friends mutually cited by both husband and wife underscored the importance of the husband in determining primary relations of the couple. (This conclusion was even more strongly supported if friends cited only by the husband were taken and considered together with mutually agreed upon husband-wife friendships.) A composite picture of the friendship choices for the three samples classified according to who initiated the friendship, whether such friendship units were local or nonlocal, a couple, or single individual is presented in table 24.1.

An examination of the table (regardless of whether the friends are local or non-local) suggests that a broad pattern of agreement exists among the groups with respect to who initiates the friendship. There are, however, two differences among the three samples. First, the wives in Sample C initiated mutually cited friendships for the pair much more often than the wives in Sample A. Also in Samples B and C, there is a greater proportion of mutually cited friends for which neither spouse can be credited with having initiated the friendship. Second, there is a greater proportion of couples in the friendship units and a smaller proportion of single individuals in Sample C as compared with Sample B, and in Sample B as compared with Sample A. Both differences (wives initiating friendships and fewer single persons as primary friends) can be explained in part by how long the couple has been married. Before and shortly after marriage single individuals predominate in the network of primary friends of both spouses. However, after marriage friends of the husband are more likely to become friends of the pair than are single friends of the wife. Indeed, the woman orients herself away from her own primary friends during the period of engagement. This shift begins during courtship, a period in which the male more often determines the kinds of activities the couple will engage in, where the couple will go, and more importantly for the present analysis, the people with whom the couple will as-

Table 24.1. Summary of Friendship Units Reported by 117 Couples According to Who Initiated Friendship

Spouse Initiating Friendship		Units on Which Couples Agree			Listed by Husband But not by Wife			Listed by Wife But not Husband		
		Couple	Single	Total	Couple	Single	Total	Couple	Single	Total
Local Friends	Sample A — Husband	34	25	59	20	20	40	6	8	14
	Neither	6	3	9	2	1	3	5	1	6
	Wife	9	9	18	6	4	10	11	9	20
	Sample B — Husband	35	10	45	18	12	30	7	5	12
	Neither	10	1	11	1	1	2	10	2	12
	Wife	10	4	11	5	3	8	8	2	10
	Sample C — Husband	33	2	35	34	6	40	18	—	18
	Neither	11	—	11	9	—	9	11	—	11
	Wife	22	1	23	10	2	12	14	9	23
Nonlocal Friends	Sample A — Husband	10	8	18	7	17	24	3	4	7
	Neither	4	—	4	4	2	6	2	1	3
	Wife	3	7	10	4	—	4	13	9	22
	Sample B — Husband	30	10	40	10	5	15	11	2	13
	Neither	5	3	8	2	1	3	3	1	4
	Wife	14	1	15	5	3	8	10	7	17
	Sample — Husband	22	2	24	15	8	23	19	5	24
	Neither	10	1	11	5	—	5	15	1	16
	Wife	11	1	12	4	—	4	15	7	22

sociate. Male superordination continues for a long as the couple is married but is more pronounced during the early years of marriage. Early in the family cycle, wives are especially accommodating. At this time, many single individuals who are close to the pair are themselves eminently marriageable; during the course of interviewing it was often found that such friends were engaged to be married and a close relationship was already developing between the fiance and the married couples whom we were questioning. Such friendships were usually male initiated. The data on Sample B and to a greater extent on Sample C revealed that by the time a couple had been married three years, almost all of their mutual friends were married and that they visited one another and formed very close friendships as couples. Also by then, the couple was meeting many more persons together, some of whom became primary friends, and wives were exercising some initiative in developing contacts for the pair.

Respondents who listed married persons as primary friends, with few exceptions, included the spouses of such persons as primary friends. Yet, the interviews showed that respondents were not always equally close to both individuals. Inclusion of both spouses as primary friends stemmed from the pattern of visiting, which most often took place on a couple basis. Thus, males who were very close usually interacted in settings where their wives were present and consequently the wives were also viewed as very close. Most of the time, husbands and wives saw themselves as a unit vis-à-vis other friends and stated that their friends regarded them as a unit.

In discussing ties with other couples, respondents often implied that they were equally close to both persons early in the interview. Yet, it became apparent in a majority of cases that respondents interacted more frequently and had closer ties with persons of the same sex. In other words, husbands were closer friends of other husbands and wives were closer to one another than they were to other husbands. Many respondents stated explicitly that it would not be morally appropriate to be equally close to both married persons who were mutual friends, and a majority of subjects supported this position indirectly.

Friends and Relatives

Support was not found for the hypothesis that couples who visited relatives frequently would interact less with primary friends. Nor did such couples have fewer primary friends. Indeed, there were a number of couples who saw relatives frequently and who also maintained a very active life with primary friends; others who saw either relatives or friends quite often but not both; and those who did not visit often with relatives or friends. Frequent association with relatives did not predict what the pattern of visiting with friends would be. Similarly, getting together frequently with friends did not appear to bear any relationship to how often couples saw relatives. Undoubtedly, some of the relatives who were visited regularly would not have been viewed as primary relations by either husband or wife, or both. . . .

Primary Friends Independent of Spouse

It was expected that the friendships a couple shared in common whould be related to the number of friends the spouses had independent of each other. First, because of the limitation imposed by time and the problems associated with maintaining such relationships and second, because the need for being on intimate friendship terms could be satisfied by having friends in common with the spouse. Specifically, the hypothesis proposed that spouses who have close friends apart from those shared would, as a result, have fewer friends in common. The test of this hypothesis was limited to the data collected in the present inquiry, that is, to the couples in Samples B and C. Respondents were asked to list all of their close friends independent of their spouse and to indicate how often they saw such persons. This information was sought separately from husband and wife, and those who were mentioned by them were not discussed in the joint interview. In 25 cases (13 in Sample B and 12 in Sample C), both husband and wife stated that they did not have any such friendships—that all of their very close friends were also close to the spouse. In an additional 29 cases, only one of the two spouses named a person or persons who were not mutual friends of the pair. Finally there were 12 couples in each of the two samples where both spouses said they had primary friends who were not close friends of their mate. The extent to which husbands and wives had friends independent of one another is presented in table 24.2. The data did not support the hypothesis. Indeed, in the random sample, there was a tendency for couples with three or more mutual friends also to have friends independent of one another, and for couples with two friends or less not to list any friends they did not share with their spouses. No such clear trend was discernable in Sample B.

Table 24.2. Nature of Contact with Primary Friends Whom Husband and Wife Have Apart from One Another

	Sample B		Sample C	
	Primary Friend of		Primary Friend of	
Nature of Contact	Husband	Wife*	Husband	Wife**
Frequent (almost daily as neighbor or work colleague)	20%	26%	56%	34%
Frequent (weekly to once in two months—not as neighbor or work colleague)	40	27	14	33
Infrequent (see less than six times per year or only correspond with)	40	47	30	33
	100%	100%	100%	100%
	(n = 42)	(n = 51)	(n = 77)	(n = 46)

* Data incomplete for six persons listed by wife.
**Data incomplete for two persons listed by wife.

Contacts by either mate with their own "private" friends were grouped in three categories. The first included persons seen almost daily as colleagues in the work environment or in the case of home-makers, as neighbors. A second category included those who were seen as often as two or three times a week but not less than once in two months. And the third category included persons whom the respondents saw less than six times per year; in fact, several persons in this category only corresponded with the respondents. The proportion of friends in these three categories is presented in table 24.2.

In general, the data suggest that middle-class couples do not maintain extensive friendship networks independently. Indeed, approximately half of the husbands and wives in our study claimed not to have a single primary friend independent of their spouse. Subjects occasionally stated that work colleagues or neighbors with whom they were on close terms did not get along well with their spouses or that they themselves had difficulty in developing a tie with the friend's spouse. An effort was usually made to establish a close relation on a couple basis, but this effort did not always succeed.

Children and the Primary Associations of Couples

One of several factors that it was felt might bear on the friendship patterns of adults was whether or not the couple had children.

Young children might make visiting difficult and inhibit parents from cultivating friendships. But, on the other hand, couples might become acquainted and develop close ties because of their children.

Though respondents readily provided information about their experience with and thinking about the role played by children, their own and their friends', they often registered surprise that children might be considered a significant factor in this sphere. However, with the exception of two friendship units, children were not instrumental in initiating contacts for adults. Furthermore, they did not appear to play a constraining role nor enhance the prospects that their parents would interact more often with primary friends.

The interviews suggest several reasons why children do not enter into the picture. During the early married years, there is a period when there are no children and when the couple is willing to go out of its way to visit with friends who have young children. When children come, apparently it is easy enough to have friends come over after the children have been bedded down or to get someone to baby-sit when the couple wants to go out. Then, too, couples with infants and young children occasionally visit each other and bring their children with them. In this regard, middle-class nuclear families tend toward equalitarian relations among all members, but parents can hardly be expected to consult their young children about taking them to visit friends. Couples with a number of children, of course, soon enough reach the point where they rely on the older children to take care of the younger ones. And couples with older children or grown children do not concern themselves with this issue.

Putting it another way, young children may have constituted an impediment for their parents when it came to friendships, but the parents did not view this as a serious matter. They were willing to make the extra effort required to maintain the close ties they already had or to make the necessary effort to establish a friendship that looked promising. For the most part, parents adopted the view that the time period in which children presented a problem would be of short duration; data on the family life cycle support them in this view.

Friends in the Local Community

The number of friends couples have in the local community does not vary with the length of time married but reaches a plateau early in marriage and remains there. Thus, couples in their forties and fifties who have been married for a long time are not likely to have more local primary friends than couples who are younger and married for a short time. When couples list many mutual friends (e.g., six or seven), a number of these are ordinarily residing in other communities; consequently, couples, regardless of age, listing as many as three or four friendships in the local community are an exception. The modal number of local primary friend units in the present study was two; indeed, three out of four couples cited no more than two units as primary in the community in which the lived.

Spouses occasionally agreed that persons with whom they had only recently become acquainted were likely to become primary friends. We were not able to explore systematically why they felt that way. Couples did not have difficulty in recalling the dynamics of their friendship relations over a long period, and it probably would not have been difficult to obtain a profile on all friendships of couples whatever the length of time they had been married. Primary friends couples acquire as adults are unlikely to be forgotten even when such persons do not keep in touch or when contact with them is lost. Indeed, primary friendships develop in adulthood between persons of roughly the same age. Our data indicated that rarely did individuals have primary friends whom they met as children in the neighborhood setting and with whom they continued to associate into adulthood and maturity. Such a notion is mainly fiction.

Why the number of local friends remained more or less constant was not clear. A pattern was observed whereby couples would sometimes become less close to one unit and substitute another; there was no steady increment in number.

DISCUSSION

This inquiry empirically supports the contention that couples constitute the basic unit with respect to primary friendship relations in the middle class in our society. There is a tendency not only for spouses to see themselves as a unit in the friendship network but for their friends to see them as a unit. In some cases, spouses maintain primary friends apart from one another. But often such friendship ties are attenuated by lack of frequent contact or by the fact that association takes place only under special conditions as, for example in the work setting. Individuals usually attempt to incorporate friends they do not share with spouses into mutual friendships; they do not always succeed.

Although the era in which we live is witnessing a decline in the dominant position of the male, our study indicates that he still enjoys considerable predominance. This predominance is all the more important because our subjects were middle class—a class in which women tend toward equality with men in most areas. Husbands are likely to initiate friendships for the pair early in marriage and continue to exercise greater influence throughout the marriage. However, as the marriage progresses, wives exercise somewhat greater influence. The longer the couples has been married the more likely the spouses are to meet persons at the same time who eventually become primary friends. But the fact remains that males are dominant at all stages in the marriage.

The contribution made by relatives to the total resources of couples' primary experiences seems to be independent of the contribution made by friends. This interpretation is supported by the fact that interaction with relatives does not appear to be quantitatively related to interaction with friends and vice versa. The schedule used did not seek information on the specific activities couples engaged in with relatives as compared with friends nor assess the affectional ties that existed between spouses and relatives. But the finding that the frequency of interaction with relatives is not related to interaction with friends, together with certain cues provided during the interview, gives the impression that orientation toward activity and affect is quite different between spouses and relatives as compared with spouses and friends. Such differences may be a function of having at one time shared a "community of fate" with one's relatives. Or it may be a function of the daily association that took place over a period of years by the spouses with the members of their family of orientation. Although relations between primary friends may be viewed uncritically, be spontaneous, be characterized by positive affect, etc., nevertheless they have lacked the constancy and "inevitability" of interaction that distinguishes relations in the nuclear family. This may result in qualitative differences in the primariness of relations with friends as compared with primariness with relatives.

There were substantial differences in style of association (measured by frequency of contact, types of activities engaged in, etc.) between couples and their friends. However, the actual number of primary friends in the local community was not great. Couples who listed a number of friends (which represented an average) were often curious to know how their number compared with that of other couples, and it sometimes surprised them that a small number was the rule. But despite the apparent lack of numbers, respondents gave the impression that they had sufficient primary-group resources at their disposal. Perhaps they did. In our present state of knowledge it is too early to tell what constitutes an adequate pattern of primary relations to enable a person to lead a well-balanced and healthy life. Lack of "adequate" association may result in or contribute to serious personal maladjustment. But what constitutes adequate association? It may well be that only two or three primary friends are all that is necessary. Or possibly one can function quite effectively with a combination of a small number of primary friends and a larger number of acquaintances, or with a combination of relatives and primary friends.

REFERENCES

Babchuk, N. and A. Bates. 1963. The primary relations of middle-class couples: a study in male dominance. *American Sociological Review* 28:377–384.

Bates, A. and N. Babchuk 1961. The primary group: a reappraisal. *Sociological Quarterly* 2:181–191.

Part V

Late Adulthood
Mastery and Competencies

Twenty-Five

The Role of Grandparenthood

Ivan F. Nye and Felix M. Berardo

One way in which older adults demonstrate competence within the context of the family is through the role of grandparenthood. In this selection, Nye and Berardo summarize what is known about the meaning of grandparenting in America. They suggest that the grandparenthood role varies a great deal from family to family and, therefore, that the experience of grandparenthood may have different implications, with regard to mastery and competence, for different people.

THE LONG-TERM RISE in average life expectancy and the consequent expansion of the latter half of the family life cycle has resulted in a substantial proliferation of surviving three-generation (or more) families in the United States. Increased life expectancy, in conjunction with other changes in family structure and composition—such as the earlier age at marriage, a shorter child-rearing period, and fewer children—have exposed more middle-aged and older couples to the role of grandparenthood than at any other period in history.

Strange as it may seem, however, the sociology and social gerontology of the grandparenthood role has remained a relatively undeveloped area from the point of view of empirical investigations. Most of the evidence we do have regarding grandparenthood is apparently derived from broader sociological studies of the aged (Neugarten and Weinstein 1964). Psychological interest in the phenomena of grandparenthood has also been minimal (Goldfarb 1965).

Cross-cultural analyses by anthropologists have provided some insight into the role of grandparents. One of the better-known studies is the research by Apple, in which ethnographic data from 75 societies were compared to determine the influence of differential patterns of family authority on the social relationships between grand-parents and their grandchildren. She found that in societies in which grandparents retain considerable household authority (by virtue of their economic power and/or because the aged are traditionally re-spected and given prestige), the relationships between the grand-parents and the grandchildren are typically *not* of "friendly equality," but rather formal and authoritarian. Conversely, in societies in which the grandparents' generation retains little or no authority over the parental generation following the birth of grandchildren, the inter-action between the grandparents and the grandchildren is typically friendly and warm, and characterized by an equalitarian or indulgent relationship. On the basis of these findings, Apple enunciates as a general principle that friendly relations between grandparents and grandchildren will occur where the family structure precludes the grandparents from exercising family authority (Apple 1956).

　　Although the grandparental role has not been extensively or systematically studied in the United States, the available evidence suggests that the principle just stated is generally applicable to in-tergenerational relationships in our own society. Although American grandparents engage in a companionable and indulgent relationship with their grandchildren, their role is usually devoid of any direct responsibility and authority (Updegraff 1968). Albrecht, for example, found that grandparents had no coveted responsibility for grandchil-dren. Instead, they adhered to a hands-off policy, whereby it was understood that the authority and the responsibility for rearing and supervising grandchildren reside solely with their parents and that grandparents were not expected to interfere except in unusual cir-cumstances. "Grandparents can give direct aid or take full respon-sibility for third-generation members only if the children are orphaned, or if the parents are absolutely unable to care for them." Nevertheless, the majority of American grandparents evidently derive pleasure, pride, and an emotional satisfaction from interacting with their grand-children (1954:201–204).

STYLE OF GRANDPARENTHOOD

At a more specific level, however, there is apparently considerable variability from one family to another regarding the actual enactment of the grandparental role. An illustration of this variability is provided by Neugarten and Weinstein, who conducted interviews with 70 sets of middle-class grandparents (46 maternal, 24 paternal) residing within the metropolitan Chicago area. Through an inductive analysis of their data, the researchers were able to differentiate five major styles of grandparenthood (1964). Their descriptions of these are presented here, with the most frequently occurring style appearing first.

1. The *Formal* are those who follow what they regard as the proper and prescribed role for grandparents. Although they like to provide special treats and indulgences for the grandchild, and although they may occasionally take on a minor service such as baby-sitting, they maintain clearly demarcated lines between parenting and grandparenting, and they leave parenting strictly to the parent. They maintain a constant interest in the grandchild but are careful not to offer advice on child-rearing.
2. The *Fun-Seeker* is the grandparent whose relation to the grandchild is characterized by informality and playfulness. He joins the child in specific activities for the specific purpose of having fun, somewhat as if he were the child's playmate. Grandchildren are viewed as a source of leisure activity, as an item of "consumption" rather than "production," or as a source of self-indulgence. The relationship is one in which authority lines—either with the grandchild or with the parent—are irrelevant. The emphasis here is on mutuality of satisfaction rather than on providing treats for the grandchild. Mutuality imposes a latent demand that both parties derive fun from the relationship.
3. The *Distant Figure* is the grandparent who emerges from the shadows on holidays and on special ritual occasions such as Christmas and birthdays. Contact with the grandchild is fleeting and infrequent, a fact which distinguishes this style from the *Formal*. This grandparent is benevolent in stance but essentially distant and remote from the child's life, a somewhat intermittent St. Nicholas.
4. The *Surrogate Parent* occurs only, as might have been anticipated, for grandmothers in this group. It comes about by initiation on the part of the younger generation, that is, when the young mother works and the grandmother assumes the actual caretaking responsibility for the child.

5. The *Reservoir of Family Wisdom* represents a distinctly authoritarian patricentered relationship in which the grandparent—on those rare occasions when it occurs in this sample, it is the grandfather—is the dispenser of special skills or resources. Lines of authority are distinct, and the young parents maintain and emphasize their subordinate positions, sometimes with and sometimes without resentment.

Approximately 32 percent of all the grandparents in this study adopted the formal style, 26 percent the fun-seeking style, 24 percent the distant-figure style, 7 percent the parent-surrogate style, and 4 percent the reservoir-of-family-wisdom style.

Neugarten and Weinstein also found that the fun-seeking and the distant-figure grandparental styles were significantly more characteristic of *middle-aged* grandparents, whereas the formal style was more likely to be adopted by *older* grandparents. These differences were true for both grandmothers and grandfathers.

GRANDMOTHERHOOD AND GRANDFATHERHOOD

There are some apparent and distinct differences between the sexes in our culture with respect to the anticipation of and/or preparation for and reactions to the grandparental role. Moreover, the magnitude of these differences varies according to whether this role is achieved relatively early in the life cycle or in later years.

Both psychological and sociological data suggest that women are much more likely than men to undergo *anticipatory socialization* with respect to the grandparental role. This involves, among other things, periodically visualizing themselves as grandmother *prior* to the actual birth of grandchildren and often before their adult children are even married. Through the process of anticipatory socialization, mothers are able to rehearse the grandparental role by developing a grandmother self-image (Cavan 1962). Typically this image is positive in nature and one that most women desire to give eventual expression to upon the arrival of grandchildren. Among some mothers, however, the grandparental role is perceived and experienced with mixed sentiments and anxiety (Deutsch 1945). Thus, a forty-two-year-old mother who perceives herself as youthful and attractive may view grandmotherhood as a threat to this self-image and therefore resent the new role or in some instances attempt to ignore or reject it entirely.

The more common reaction among middle-aged mothers, however, is probably a combination of joy, excitement, and pride. Persons achieving grandparental status in our culture usually gain additional respect and prestige in the eyes of other members of society. Moreover, as Leslie has noted

grandmotherhood often is a major part of the middle-aged woman's solution to the loss of her children through marriage. As grandmother, she acquires a new sense of importance and usefulness. Moreover, she experiences again most of the joys of parenthood without having to cope with the exacting demands. Entering her children's homes as a visitor and/or baby-sitter, she can indulge herself and her grandchildren; when her energy or her patience wanes, she has simply to leave and go to her own quiet home (1967:684–686).

The grandmotherly pattern established in middle age is easily carried over and maintained in old age unless the physical disabilities that often accompany the advanced years intrude to prevent its continuation.

Middle-aged fathers who find themselves grandfathers are less likely to be concerned about their new role. For most men the middle years represent the period when they are reaching the apex of their occupational careers and economic success. They are still primarily identified with and engaged in the work role. Consequently, intense male involvement in the grandfatherly role typically is postponed until the years following retirement. It is only at this later date that grandparental activities are apt to become a major focus of attention for men.

It has been suggested that in contemporary American society a general cultural definition of the elderly grandfather role has emerged that is essentially maternal in nature (Cavan 1962). Unlike his counterpart in the patriarchal family of the past, the modern grandfather does not function as a primary source of authority for his grandchildren and, except under unusual circumstances, he is neither expected nor permitted to act as their financial provider. Instead, he assumes a slightly masculinized grandmother role involving such maternal tasks as feeding the grandchildren and baby-sitting for them, taking them for [carriage] rides, and so on. To make a successful accommodation to this new role, the grandfather must develop an orientation that differs considerably from the masculine instrumental role he has performed most of his adult life. Some make the transition smoothly; others do so with some difficulty because the maternal quality of the behavior required is either repugnant or embarrassing to them. The culture provides assistance here by attaching status and respect to the grandfather role. Men occupying this position are generally the recipients of verbal compliments and praise from other members of society. The American grandfather is encouraged and permitted to engage in a quasimaternal relationship with his grandchildren without discomfort or embarrassment.

Increasing numbers of Americans are moving into the grandparental role. Because this role typically entails a minimum of obligations and responsibilities, most persons find that its enactment allows a variety of opportunities for personal gratification and self-enhancement. Moreover, the grandparent functions to provide continuity in family interaction, and this function in itself gives persons who fulfill this role a valued position in American society.

REFERENCES

Albrecht, R. 1954. The parental responsibilities of grandparents. *Marriage and Family Living* 16:201–204.

Apple, D. 1956. The social structure of grandparenthood. *American Anthropologist* 58:656–663.

Cavan, R. 1962. Self and role in adjustment during old age. In A. Rose, ed., *Human Behavior and Social Process*, pp. 526–535. Boston: Houghton Mifflin.

Deutsch, H. 1945. *The Psychology of Women*, vol. 2. New York: Grune and Stratton.

Goldfarb, A. 1965. Psychodynamics and the three-generation family. In E. Shanas and G. Streib, eds., *Social Structure and the Family: Generational Relations*, pp. 10–45. Englewood Cliffs, N.J.: Prentice-Hall.

Leslie, G. 1967. *The Family in Social Context*. New York: Oxford University Press.

Neugarten, B. and K. Weinstein. 1964. The changing American grandparent. *Journal of Marriage and the Family* 26:199–204.

Updegraff, S. 1968. Changing role of the grandmother. *Journal of Home Economics* 60:177–180.

Twenty-Six

Adjustment to Loss of Job at Retirement
Robert C. Atchley

Adjusting to retirement is a major event in the domain of mastery and competence for many older adults. As Robert Atchley suggests, however, it is difficult to generalize about the effects of retirement on individuals. How an individual adjusts to retirement depends primarily on what the meaning of work was to that person throughout adulthood (in this regard, see selection 20), on the individual's general capacity to cope with change, and on the individual's view of the retirement role.

HOW DO PEOPLE adjust to retirement? This is an area in which there has been a fair amount of systematic research as well as a good deal of theorizing. . . . Theory is a necessary element in the development of knowledge. Theory serves to identify gaps in knowledge and as a guide for organizing research. With regard to adjustment to retirement, theories are needed to describe and explain adjustment to changes in income and adjustment to no longer having job responsibilities.

As yet no theory concerning how people adjust to income reduction has been developed. What is needed is descriptive data on the phases of adjustment and on the strategies that are employed. A decision model would probably be useful. However, at this point the basic descriptive data necessary to begin to develop such a theory have not been assembled. More attention has been given to the problem of adjusting to the loss of one's job. Several theories of adjustment have emerged, each of which has a different emphasis.

Activity theory assumes that the job means different things to different people and that to adjust successfully to the loss of one's job, one must find a substitute for whatever personal goal the job was used to achieve. The most often quoted proponents of this theory are

331

Friedmann and Havighurst (1954) and Miller (1965). Friedmann and Havighurst approached the matter in terms of substitute activities, and Miller carried it one step further to include substitute activities which serve as new sources of identity. The assumption here is that the individual will seek and find a work substitute. In a test of this theory, however, Shanas (1972) found it to be of very limited utility when applied to American society. In my own research, activity theory has fit the behavior of only a tiny proportion of retired people (Atchley 1971).

Continuity theory assumes that, whenever possible, the individual will cope with retirement by increasing the time spent in roles he already plays rather than by finding new roles to play (Atchley 1972). This assumption is based on the finding that older people tend to stick with tried and true ways rather than to experiment and on the belief that most retired people want their life in retirement to be as much like their preretirement life as possible. However, continuity theory allows for a gradual reduction in overall activity. Obviously, these assumptions do not fit *all* retired people, although they may fit the majority.

Disengagement theory (Cumming and Henry 1961; Cumming 1964; Henry 1965) holds that retirement is a necessary manifestation of the mutual withdrawal of society and the older individual from one another as a consequence of the increased prospect of biological failure in the individual organism. This theory has been criticized for making the rejection of older people by society seem "natural" and, therefore, right. However, the fact of the matter is that many people *do* want to withdraw from full-time jobs and welcome the opportunity to do so. Streib and Schneider (1971) refined disengagement theory to apply more directly to the realities of retirement. *Differential disengagement* is the term they use to reflect the idea that disengagement can occur at different rates for different roles. And they do not contend that disengagement is irreversible. They hold that by removing the necessity for energy-sapping labor on a job, retirement may free the individual with declining energy to increase his level of engagement in other spheres of life. Such increases in involvement do in fact occur (Cottrell and Atchley 1969).

Thus, both continuity theory and the theory of differential disengagement attempt to explain why people do not adjust to retirement as activity theory would lead us to expect.

Hazardous as it may be, I would like to offer another approach to understanding adjustment to the loss of job that accompanies retirement. It is to some extent speculative, but I think it is also based on the facts as we now know them. It also synthesizes the major elements of the three theories mentioned above. In sum, the central processes of adjustment are held to be *internal compromise* and *interpersonal negotiations.* While these two elements interact quite strongly in real situations, for analytical purposes I shall treat them separately.

When a person retires, a new role is taken on and an old one relinquished, at least to a degree. The extent to which this triggers a need for a new adjustment on the part of the individual depends on how the job role fits into his pattern of adjustment prior to retirement.

It is probably safe to say that retirement represents a certain amount of disruption in the lives of just about everyone who retires. But why is this disruption so much more serious for some than for others? The answer to this is to be found in examining the relationship between the amount of change introduced by retirement and the capacity of the individual to deal with change routinely.

When people can deal with a substantial amount of change in a more or less routine fashion, we call them flexible. We also tend to think of changes as *serious* only if they exceed the magnitude that can be dealt with routinely by the individual. People who have difficulty adjusting to retirement can be people with a low level of tolerance for any change (inflexible). They can also be people who are confronted with especially serious change. Obviously, the degree to which any change is serious depends on how adaptable the individual is. Thus, among those who have difficulty adjusting to retirement, we would expect to find a group of rigid, inflexible people for whom even small changes in the status quo are seen as serious. We would also expect to find a group of reasonably flexible people who are having to adjust to what seems *to them* a high magnitude of change.

A second important variable is the individual's hierarchy of personal goals. Everyone has personal goals—important results which, if achieved, give the individual a strong sense of personal worth or satisfaction. These goals are of several types. Some of them involve learning to respond to life in a particular manner—the development of certain personal qualities such as honesty, ambition, cheerfulness, kindness, and so on, which can be exhibited in most situations and in most roles. Other personal goals are materialistic and involve achieving ownership of particular property such as land, house, stereo, car, etc. Still other personal goals involve successfully playing certain roles. Thus, a person's desire to succeed as a parent, a job-holder, an artist, or any number of other roles can also be viewed in relation to this hierarchy of personal goals. . . .

An individual's personal goals are organized into a hierarchy which indicates the individual's priorities for achieving them. This hierarchy reflects the *relative* importance of particular personal goals. The hierarchy and the personal goals that comprise it change constantly as goals are added or dropped and as succcess or failure alters priorities. Personal goals come from three major sources: goals we are taught and are expected to hold as personal goals; personal goals that are held by others we seek to emulate; and personal goals that grow out of our own experiences and knowledge about ourselves and our capabilities. Any or all three sources in various combinations may contribute to motivation toward a given personal goal.

An important dimension of the norms concerning the individual's hierarchy of personal goals involves what is supposed to happen in retirement. The norms demand that upon retirement, the job is no longer eligible to occupy a top spot in the hierarchy of personal goals, and factors such as managing one's own affairs and maintaining an independent household are moved up on the list of expected priorities. . . .

For our purposes, the most important compromises involve the top priority *roles* in the hierarchy. The individual starts off in adolescence with a tentative hierarchy. To the extent that he can gain experience through school, summer jobs, or other means concerning how workable the tentative hierarchy of roles is, the individual can gradually move toward stabilizing the hierarchy, particularly the top-priority roles. This sort of stability is important at this stage because it gives the budding adult the security of a firm sense of purpose—of knowing where he is going. Of course, achieving this sort of stability may take some time and may be quite painful. Some people never achieve it, others may achieve stability early and stick with the same hierarchy their entire lives, others may experience several stable hierarchies over their lifetime, and still others may experience stable hierarchies over their lifetime, and still others may gravitate in and out of a feeling of stability.

Such stability in the hierarchy of personal goals may be achieved in two ways. The hierarchy may be positively reinforced, become satisfying, meet with success, and, therefore, attract the individual's commitment. Or the hierarchy may produce results that are neither bad enough to cause it to be abandoned nor good enough to attract the individual's commitment, but at the same time the hierarchy may have been used long enough to have become a habit. For people who are *committed* to a hierarchy with the job at or near the top, retirement is more difficult than for people for whom the job is a prime consideration only out of habit. Of course, retirement is but one of many changes that *can* destroy the stability of one's hierarchy of personal goals. Whether retirement *does* destory this stability depends on where the job role fits into this hierarchy.

For people who stress personal qualities, the job may be quite far down on the list. Materialistic people may consider the job more important—to the extent that it governs their ability to achieve materialistic goals. Even among people who judge their successes primarily in terms of role performance, however, the job may not be a primary source of satisfaction. . . .

The process involved in developing or changing the hierarchy of personal goals is, of course, decision-making, and the expression *internal compromise* is used to describe this decision-making process in order to indicate that its outcome is far from determinate. The process whereby people reorganize their criteria for decision-making, though it may exist, is one we know precious little about.

An important aspect of all this is just where the retired role (as opposed to the job role) fits into this hierarchy. Some people may resist including successful retirement in their hierarchy of personal goals at all. Still others may put it at the bottom. Increasingly, however, playing the role of retired person seems to be taking a high position in the hierarchy and, therefore, can be expected to play a part in the reorganization of the hierarchy. Research is needed to establish just where the retirement role fits into the structure of personal goals and under what circumstances its rank may vary.

Interpersonal negotiations is a process that articulates the individual's goals and aspirations with those of the people he interacts with. It is through this process that the world outside the individual can influence development of and change in his hierarchy of personal goals. When we say that we "know" a person, one of things we "know" is his hierarchy of personal goals. When this hierarchy changes, the individual indicates to others, through his decisions or through his actions, that a change has taken place. I use the term *negotiation* here because often the individual runs into resistance in getting others who are important to him to accept his new hierarchy of personal goals. And at this point, the results of internal compromise and feedback from significant others enter into a dialectic.

Unfortunately, while the interpersonal negotiation process is no doubt important to the development of a stable hierarchy of personal goals in retirement, research evidence provides few clues as to how this process works and how retirement and job roles are dealt with in the process.

SUMMARY

Not quite a third of the retired population encounters difficulty in adjusting to retirement. Getting used to reduced income is by far the most frequent reason for such difficulty (40 percent). Missing one's job accounts for about 22 percent of the adjustment difficulties. The remaining 38 percent is accounted for by factors such as death of spouse or declining health, which are directly related to retirement adjustment only in that they influence the situation in which such adjustment must be carried out. This suggests that certain situational prerequisites are often necessary to a good adjustment.

From a positive point of view, it seems safe to say that adjustment to retirement is greatly enhanced by sufficient income, the ability to give up one's job gracefully, and good health. In addition, adjustment seems to be smoothest when situational changes other than loss of job are at a minimum. Another way of viewing this is to say, assuming that one's fantasy about the retirement role is based on reality, that factors which upset the ability of the retirer to live out his retirement ambitions hinder his ability to adjust to retirement smoothly.

People who have difficulty adjusting to retirement tend to be those who are either very inflexible in the face of change or faced with substantial change or both. The prime things about retirement that must be adjusted to are loss of income and loss of job. We know very little about how people adjust to loss of income.

A theory of how people adjust to a loss of job at retirement was presented which attempts to integrate existing theories by means of the impact of retirement on the individual's hierarchy of personal goals. If the job is high in that hierarchy and yet unachieved, then the individual can be expected to seek another job. If a job substitute cannot be found, then the hierarchy of personal goals must be reorganized. If the individual is broadly engaged, then he must search for alternate roles. If he is successful, he then develops a new hierarchy of personal goals. If not, he withdraws. Of course, if the job is not high in the hierarchy to begin with, it requires no serious change in the personal goals that form the basis for everyday decisions.

REFERENCES

Atchley, Robert C. 1971. Retirement and leisure participation: continuity or crisis? *Gerontologist* 11:1(1):13–17.
—— 1972. *The Social Forces in Later Life: An Introduction to Social Gerontology*. Belmont, Calif.: Wadsworth.
Cottrell, Fred and Robert C. Atchley. 1969. *Women in Retirement: A Preliminary Report*. Oxford, Ohio: Scripps Foundation for Research in Population Problems.
Cumming, Elaine. 1964. New thoughts on the theory of disengagement. In Robert Kastenbaum, ed., *New Thoughts on Old Age*. New York: Springer.
Cumming, Elaine and William E. Henry. 1961. *Growing Old: The Process of Disengagement*. New York: Basic Books.
Friedmann, Eugene and Robert J. Havighurst, eds. 1954. *The Meaning of Work and Retirement*. Chicago: University of Chicago Press.
Henry, William E. 1965. The theory of intrinsic disengagement. In P. From Hansen, ed., *Age with a Future*. Copenhagen: Munksgaard.
Miller, Stephen J. 1965. The social dilemma of the aging leisure participant. In Arnold M. Rose and Warren A. Petersen, eds., *Older People and Their Social World*, pp. 77–92. Philadelphia: F. A. Davis.
Shanas, Ethel. 1972. Adjustment to retirement: substitution or accommodation? In Frances M. Carp, ed., *Retirement*, pp. 219–244. New York: Behavioral Publications.
Streib, Gordon F. and Clement J. Schneider. 1971. *Retirement in American Society*. Ithaca, New York: Cornell University Press.

Late Adulthood
Identity and the Self

Twenty-Seven

Personality and Patterns of Aging

Robert J. Havighurst

What is successful aging? In this selection, Robert Havighurst examines two contrasting viewpoints—*activity theory*, which suggests that optimal aging is achieved by remaining active and socially engaged as long as possible, and *disengagement theory*, which suggests that successful aging involves a withdrawal of the individual from society and vice versa. The author suggests that it is impossible to make a hard and fast generalization about which pattern of aging is superior since neither theory takes into account the individual's personality. For some persons, activity seems to work best; for others, successful aging involves a certain degree of disengagement.

HAPPINESS AND SATISFACTION in the latter part of life are within reach of the great majority of people. The external conditions of life are better for people over sixty-five than they have been at any time in this century. Social Security benefits and company pensions are at record high levels. Medicare has underwritten much of the major medical expense. Almost no one is forced to work after the age of sixty-five if he prefers not to. Most of the states have programs, supported under the Older Americans Act, to improve the social adjustment of older people. In other words, society has done just about as much as anyone could ask it to do on behalf of older people. At least this is the conclusion one would draw from a superficial look at social statistics.

This paper is a result of cooperation among the following three people: Bernice L. Neugarten, Sheldon S. Tobin, and Robert J. Havighurst. A description of the central research has been published by Dr. Neugarten under the same title in *Gawein* 13:249–256, 1965.

341

Yet we know that many people are unhappy and dissatisfied in their later years. Some of them suffer from poor health, but this is only a minority. The average person at age sixty-five will live four more years. According to research, the person who is in good health suffers very little impairment in his ability to learn, to initiate actions, to be effective in the ordinary relations of life until he is eighty-five years old or more.

Since a great many people after sixty-five have good enough health and enough income to support a life of happiness and satisfaction, we must turn to the psychologist to ask why some of these people are unhappy and dissatisfied. Have they been unhappy all their lives? Are they unhappy due to remediable present situations? Are there forms of psychotherapy or of environmental improvement that would substantially increase the number of happy and satisfied older people?

Theories of Successful Aging

There are at present two contrasting theories of successful aging. Both are unsatisfactory because they obviously do not explain all the phenomena of successful aging. Yet both have some facts to support them. The first, one that might be called the *activity theory*, implies that, except for the inevitable changes in biology and in health, older people are the same as middle-aged with essentially the same psychological and social needs. In this view. the decreased social interaction that characterizes old age results from the withdrawal by society from the aging person; and this decrease in interaction proceeds against the desires of most aging men and women. The older person who ages optimally is the person who stays active and who manages to resist the shrinkage of his social world. He maintains the activities of middle age as long as possible, and then finds substitutes for work when he is forced to retire and substitutes for friends and loved ones whom he loses by death.

In the *disengagement theory* (Cumming and Henry 1961), on the other hand, the decreased social interaction is interpreted as a process characterized by mutuality; one in which both society and the aging person withdraw, with the aging individual acceptant, perhaps even desirous, of the decreased interaction. It is suggested that the individual's withdrawal has intrinsic or developmental qualities as well as responsive ones; that social withdrawal is accompanied or preceded by increased preoccupation with the self and decreased emotional investment in persons and objects in the environment; and that, in this sense, disengagement is a natural rather than an imposed process. In this view, the older person who has a sense of psychological well-being will usually be the person who has reached a new equilibrium characterized by a greater psychological distance, altered types of relationships, and decreased social interaction with persons around him.

In order to test these two theories empirically, the data of the Kansas City Study of Adult Life were used, consisting of repeated interviews with 159 men and women aged fifty to ninety, taken over the period from 1956 through 1962. The sample at the end of the study consisted of 55 percent of the original people. Of the attrition, 27 percent had been due to death; 12 percent to geographical moves; and the rest to refusal to be interviewed at some time, usually because of reported poor health. There is evidence also that those who were relatively socially isolated constituted a disproportionate number of the dropouts. The original sample excluded people living in institutions and those who were so ill that they could not be interviewed. The original sample also excluded people at the very bottom of the socio-economic scale and a few who would have been diagnosed as neurotic by a psychiatrist, as well as those who were chronically ill if the illness confined them to bed. Some of the sample became quite ill, physically or mentally, during the period of the study, but they continued in the study if they could be interviewed.

The results of this study indicated that neither the activity theory nor the disengagement theory was adequate to account for the observed facts. While there was a decreased engagement in the common social roles related to increasing age, some of those who remained active and engaged showed a high degree of satisfaction. On the whole, those who were most active at the older ages were happier, but there were many exceptions to this rule.

Need for a Personality Dimension

Since it is an empirical fact that some people are satisfied with disengagement while others prefer a high degree of social engagement, it is clear that something more is needed to give us a useful theory of successful aging. Possibly that something is a theory of the relationship of personality to successful aging (Havighurst, Neugarten, and Tobin 1964; Neugarten 1965).

A substantial beginning on such a theory was made by Else Frenkel-Brunswik and her colleagues Reichard, Livson, and Petersen (1962) in their study of 87 elderly working men in the San Francisco area, 42 of them retired and 45 not retired. After interviewing these men intensively and rating them on 115 personality variables, the researchers rated them on "adjustment to aging" using a 5-point rating scale. Sixty men were rated either high (4 or 5) or low (1 or 2). Their personality ratings were subjected to a "cluster analysis" to identify men who were highly similar to one another. The high group produced three clusters and the low group produced two clusters, leaving 23 of the 60 not in any cluster. The five clusters or "types" of men were given the following names:

High on Adjustment	N	Low on Adjustment	N
Mature	14	Angry	16
Rocking-chair	6	Self-haters	4
Armored	7		

Among those judged successful in aging, the "mature" group took a constructive rather than an impulsive or a defensive approach to life. The "rocking-chair" group tended to take life easy and to depend on others. The "armored" men were active in defending themselves from becoming dependent. They avoided retirement if possible, and one of them who was ill complained of his enforced idleness, something the rocking-chair type would have been glad to accept. Even the oldest of this group, an eighty-three-year old, still worked a half-day every day.

Among those judged unsuccessful in aging, the "angry" men were generally hostile toward the world and blamed others when anything went wrong. They were poorly adjusted to work and several had been downwardly mobile socially. They tended to resent their wives. This group was especially fearful of death.

The "self-haters" differed from the "angry" men primarily by openly rejecting themselves and blaming themselves for their failures. They were depressed. Death for them was a longed-for release from an intolerable existence.

These types of men were making quite different behavioral

adjustments to aging. Thus, the armored and the rocking-chair were judged to be equally successful in adjusting to aging, but their adjustments were diametrically opposed. One group was active while the other was disengaged.

THE KANSAS CITY STUDY OF ADULT LIFE

The Kansas City Study of Adult Life carried on this search for a personality dimension by studying women as well as men, over a social class range from upper-middle class through upper-working class. The 159 persons were rated on fourty-five personality variables reflecting both the cognitive and affective aspects of personality. Types of personality were extracted from the data by means of factor analysis. There were four major types, which we have called the integrated, armored-defended, passive-dependent, and unintegrated personalities.

Patterns of behavior were defined on the basis of a rating of *activity* in eleven common social roles: worker, parent, grandparent, kin-group member, spouse, homemaker, citizen, friend, neighbor, club and association member, church member. Ratings were made by judges on each of the eleven roles, based on a reading of the seven interviews with each person. The sum of the role-activity scores was used to divide the respondents into activity levels—high, medium, and low.

A third component of the patterns of aging was a measure of *life satisfaction* or psychological well-being, which was a composite rating based on five scales recording the extent to which a person (a) finds gratification in the activities of his every-day life; (b) regards his life as meaningful and accepts both the good and the bad in it; (c) feels that he has succeeded in achieving his major goals; (d) has a positive image of himself; and (e) maintains happy and optimistic moods and attitudes. Scores on life satisfaction were grouped into high, medium, and low categories.

The analysis based on these three dimensions (personality, role activity, and life satisfaction) was applied to the 59 men and women in the study who were aged seventy to seventy-nine. This is the group in which the transition from middle age to old age has presumably been accomplished. Fifty of these people were clearly in one or another of eight patterns of aging, which are presented in table 27.1.

Group A, called the *reorganizers*, are competent people engaged in a wide variety of activities. They are the optimal agers in terms of the American ideal of "keeping active, staying young." They reorganize their lives to substitute new activities for lost ones.

Group B are called the *focused*. They are well-integrated personalities with medium levels of activity. They tend to be selective about their activities, devoting their time and energy to gaining satisfaction in one or two role areas.

Group C we call the *successful disengaged*. They have low activity levels with high life satisfaction. They have voluntarily moved away from role commitments as they have grown older. They have high feelings of self-regard, with a contented "rocking-chair" position in life.

Group D exhibits the *holding-on* pattern. They hold as long as possible to the activities of middle age. As long as they are successful in this, they have high life satisfaction.

Group E are *constricted*. They have reduced their role activity presumably as a defense against aging. They constrict their social interactions and maintain a medium to high level of satisfaction. They differ from the *foused* group in having less integrated personalities.

Group F are *succorance-seeking*. They are successful in getting emotional support from others and thus they maintain a medium level of role activity and of life satisfaction.

Group G are *apathetic*. They have low role activity combined with medium or low life satisfaction. Presumably, they are people who have never given much to life and never expected much.

Table 27.1. Personality Patterns in Aging

Personality Type	Role Activity	Life Satisfaction	N
A. Integrated (reorganizers)	High	High	9
B. Integrated (focused)	Medium	High	5
C. Integrated (disengaged)	Low	High	3
D. Armored-defended (holding on)	High or medium	High	11
E. Armored-defended (constricted)	Low or medium	High or medium	4
F. Passive-dependent (succorance-seeking)	High or medium	High or medium	6
G. Passive-dependent (apathetic)	Low	Medium or low	5
H. Unintegrated (disorganized)	Low	Medium or low	7

Group H are *disorganized*. They have deteriorated thought processes and poor control over their emotions. They barely maintain themselves in the community and have low or, at the most, medium life satisfaction.

These eight patterns of aging probably are established and predictable by middle age, although we do not have longitudinal studies to prove this proposition. It seems reasonable to suppose that a person's underlying personality needs become consonant with his overt behavior patterns in a social environment that permits wide variation.

CONCLUSIONS

In some ways the Kansas City Study and other studies of behavior and life satisfaction support the activity theory of optimal aging; as the level of activity decreases, so also do the individual's feelings of contentment regarding his present activity. The usual relationships are high activity with positive affect, and low activity with negative affect. This relationship does not decrease after age seventy.

At the same time, the data in some ways support the disengagement theory of optimal aging: There are persons who are relatively high in role activity who would prefer to become more disengaged from their obligations; there are also persons who enjoy relatively inactive lives.

Neither the activity theory nor the disengagement theory of optimal aging is itself sufficient to account for what we regard as the more inclusive description of these findings: that as men and women move beyond age seventy in modern, industrialized communities they regret the drop in role activity that occurs in their lives; at the same time, most older persons accept this drop as an inevitable accompaniment of growing old, and they succeed in maintaining a sense of self-worth and a sense of satisfaction with past and present life as a whole. Other older persons are less successful in resolving these conflicting elements—not only do they have strong negative affect regarding losses in activity but the present losses weigh heavily and are accompanied by a dissatisfaction with past and present life.

The relationships between levels of activity and life satisfaction are also influenced by personality type, particularly by the extent to which the individual remains able to integrate emotional and rational elements of his personality. Of the three dimension on which we have data—activity, satisfaction, and personality—personality seems to be the pivotal dimension in describing patterns of aging and in predicting relationships between level of activity and life satisfaction. It is for this reason, also, that neither the activity nor the disengagement theory is satisfactory, since neither deals, except peripherally, with the issue of personality differences.

REFERENCES

Cumming, E. and W. E. Henry. 1961. *Growing Old*. New York: Basic Books.
Havighurst, R. J., B. L. Neugarten, and S. S. Tobin. 1964. Disengagement, personality and life satisfaction in later years. In P. From Hansen, *Age with a Future*. Copenhagen: Munksgaard.
Neugarten B. L. 1965. Personality and patterns of aging. *Gawein* 13:249–256.
Reichard, S., F. Livson, and P. G. Peterson. 1962. *Aging and Personality*. New York: Wiley.

Twenty-Eight

The Life Review: An Interpretation of Reminiscence in the Aged

Robert N. Butler

One of the most important developmental tasks of the late adulthood years involves coming to terms with the life one had led—as Erik Erikson has termed it, achieving a sense of *ego integrity*. In this selection, Robert Butler describes a phenomenon that may be related to this ego integrity process: *the life review*. He argues that a constructive reevaluation of one's past may contribute to the development of wisdom, serenity and, to a great extent, maintenance of healthy psychological functioning.

THIS PAPER POSTULATES the universal occurrence in older people of an inner experience or mental process of reviewing one's life. I propose that this process helps account for the increased reminiscence in the aged, that it contributes to the occurrence of certain late-life disorders, particularly depression, and that it explains in part the evolution of such characteristics as candor, serenity, and wisdom among certain of the aged.

Intimations of the existence of a life review in the aged are found in psychiatric writings—notably in the emphasis upon reminiscence—and the nature, sources, and manifestations of the life review have been studied in the course of intensive psychotherapeutic relationships (Butler 1960). But often the older person is experienced as garrulous and "living in the past," and the content and significance of his reminiscence are lost or devalued. Younger therapists espe-

349

cially, working with the elderly, find great difficulties in listening (Butler 1961).

The prevailing tendency is to identify reminiscence in the aged with psychological dysfunction and thus to regard it essentially as a symptom. One source of this distorted view is the emphasis in available literature on the occurrence of reminiscence in the mentally disordered and institutionalized aged. Of course, many of the prevailing ideas and "findings" concerning the aged and aging stem primarily from the study of such samples of elderly people. Since the adequately functioning community-resident aged have only recently been systematically studied and intensive study of the mentally disturbed aged through psychotherapy has been comparatively rare (Rechtschaffen 1959), these important sources for data and theory have not yet contributed much to an understanding of the amount, prevalence, content, function, and significance of reminiscence in the aged.

Furthermore, definitions and descriptions of reminiscence—the act or process of recalling the past—indicate discrepant interpretations of its nature and function. Reminiscence is seen by some investigators as occurring beyond the older person's control: It happens to him; it is spontaneous, nonpurposive, unselective, and unbidden. Others view reminiscence as volitional and pleasurable but hint that it provides an escape. Thus, purposive reminiscence is interpreted only as helping the person to fill the void of his later life. Reminiscence is also considered to obscure the older person's awareness of the realities of the present. It is considered of dubious reliability, although, curiously, "remote memory" is held to be "preserved" longer than "recent memory." In consequence, reminiscence becomes a pejorative, suggesting preoccupation, musing, aimless wandering of the mind. In a word, reminiscence is fatuous. Occasionally, the constructive and creative aspects of reminiscence are valued and affirmed in the autobiographical accounts of famous men, but it must be concluded that the more usual view of reminiscence is a negative one.

In contrast, I conceive of the life review as a naturally occurring, universal mental process characterized by the progressive return to consciousness of past experiences, and, particularly, the resurgence of unresolved conflicts; simultaneously and normally, these revived experiences and conflicts can be surveyed and reintegrated. Presumably this process is prompted by the realization of approaching dissolution and death, and the inability to maintain one's sense of personal invulnerability. It is further shaped by contemporaneous experiences and its nature and outcome are affected by the lifelong unfolding of character.

THE SIGNIFICANCE OF DEATH

The life review mechanism, as a possible response to the biological and psychological fact of death, may play a significant role in the psychology and psychopathology of the aged. . . .

The relation of the life-review process to thoughts of death is reflected in the fact that it occurs not only in the elderly but also in younger persons who expect to die—for example, the fatally ill or the condemned. It may also be seen in the introspection of those preoccupied by death, and it is commonly held that one's life passes in review in the process of dying. . . .

But the life review is more commonly observed in the aged because of the actual nearness of life's termination—and perhaps also because during retirement not only is time available for self-reflection, but the customary defensive operation provided by work has been removed. . . .

Reviewing one's life, then, may be a general response to crises of various types, of which imminent death seems to be one instance. It is also likely that the degree to which approaching death is seen as a crisis varies as a function of individual personality. The explicit hypothesis intended here, however, is that the biological fact of approaching death, independent of—although possibly reinforced by—personal and environmental circumstances, prompts the life review.

MANIFESTATIONS OF THE LIFE REVIEW

The life review, as a looking-back process that has been set in motion by looking forward to death, potentially proceeds toward personality reorganization. Thus, the life review is not synonymous with, but includes reminiscence; it is not alone either the unbidden return of memories, or the purposive seeking of them, although both may occur.

The life review sometimes proceeds silently, without obvious manifestations. Many elderly persons, before inquiry, may be only vaguely aware of the experience as a function of their defensive structure. But alterations in defensive operations do occur. Speaking broadly, the more intense the unresolved life conflicts, the more work remains to be accomplished toward reintegration. Although the process is active, not static, the content of one's life usually unfolds slowly;[1] the process may not be completed prior to death. In its mild form, the life review is reflected in increased reminiscence, mild nostalgia, mild regret; in severe form, in anxiety, guilt, despair, and depression. In the extreme, it may involve the obsessive preoccupation of the older person with his past and may proceed to a state approximating terror and result in suicide. Thus, although I consider it to be a universal and the normative process, its varied manifestations and outcomes may include psychopathological ones.

The life review may first be observed in stray and seemingly insignificant thoughts about oneself and one's life history. These thoughts may continue to emerge in brief intermittent spurts or become essentially continuous, and they may undergo constant reintegration and reorganization at various levels of awareness. A seventy-six-year-old man said:

> My life is in the background of my mind much of the time; it cannot be any other way. Thoughts of the past play upon me; sometimes I play with them, encourage and savor them; at other times I dismiss them.

Other clues to its existence include dreams and thoughts. The dreams and nightmares of the aged, which are frequently reported (Perlin and Butler 1964), appear principally to concern the past and death. Imagery of past events and symbols of death seem frequent in waking life as well as in dreams, suggesting that the life review is a highly visual process.[2]

---

[1] The term "life review" has the disadvantage of suggesting that orderliness is characteristic. The reminiscences of an older person are not necessarily more orderly than any other aspects of his life, and he may be preoccupied at various times with particular periods of his life and not with the whole of it.

[2] Various sensory processes are involved. Older people report the revival of the sounds, tastes, smells of early life, as: "I can hear the rain against the window of my boyhood room."

Another manifestation of the life review seems to be the curious but apparently common phenomenon of mirror-gazing, illustrated by the following:

I was passing by my mirror. I noticed how old I was. My appearance, well, it prompted me to think of death—and of my past—what I hadn't done, what I had done wrong. . . .

Adaptive and Constructive Manifestations

As the past marches in review, it is surveyed, observed, and reflected upon by the ego. Reconsideration of previous experiences and their meanings occurs often with concomitant revised or expanded understanding. Such reorganization of past experience may provide a more valid picture, giving new and significant meanings to one's life; it may also prepare one for death, mitigating one's fears.[3]

Although it is not possible at present to describe in detail either the life review or the possibilities for reintegration that are suggested, it seems likely that in the majority of the elderly a substantial reorganization of the personality does occur. This may help to account for the evolution of such qualities as wisdom and serenity, long noted in some of the aged. Although a favorable, constructive, and positive end result may be enhanced by beneficial environmental circumstances, such as comparative freedom from crises and losses, it is more likely that successful reorganization is largely a function of the personality—in particular. such vaguely defined features of the personality as flexibility, resilience, and self-awareness.

In addition to the more impressive constructive aspects of the life review, certain adaptive and defensive aspects may be noted. Some of the aged have illusions of the "good past"; some fantasy the past rather than the future in the service of avoiding the realities of the present; some maintain a characteristic detachment from others and themselves. Although these mechanisms are not constructive, they do assist in maintaining a status quo of psychological functioning. . . .

[3] For example, Joyce Cary's *To Be a Pilgrim* (London: Michael Joseph, 1942) concerns an insightful old man "deep in his own dream, which is chiefly of the past," (p. 7) and describes the review of his life, augmented by the memories stimulated by his return to his boyhood home.

DISCUSSION

It is evident that there is considerable need for an intensive detailed study of aged persons in order to obtain information about their mental functioning, the experience of aging, approaching death, and dying. Behavior during aging may be clarified by the revelations of subjective experience. . . . The personal sense and meaning of the life cycle are more clearly unfolded by those who have nearly completed it. The nature of the forces shaping life, the effects of life events, the fate of neuroses and character disorders, the denouement of character itself may be studied in the older person. Recognition of the occurrence of such a vital process as the life review may help one to listen to, to tolerate, and to understand the aged, and not to treat reminiscence as devitalized and insignificant.

Of course, people of all ages review their past at various times; they look back to comprehend the forces and experiences that have shaped their lives. However, the principal concern of most people is the present, and the proportion of time younger persons spend dwelling on the past is probably a fair, although by no means definite, measure of mental health. One tends to consider the past most when prompted by current problems and crises. The past also absorbs one in attempts to avoid the realities of the present. A very similar point has been made by others in connection with the sense of identity: One is apt to consider one's identity in the face of life crisis; at other times the question of "Who and what am I?" does not arise.

One might also speculate as to whether there is any relationship between the onset of the life review and the self-prediction and occurrence of death. Another question that arises is whether the intensity of a person's preoccupation with the past might express the wish to distance himself from death by restoring the past in inner experience and fantasy. This may be related to human narcissism or sense of omnipotence, for persons and events can in this way be recreated and brought back. At the same time a constructive reevaluation of the past may facilitate a serene and dignified acceptance of death.

REFERENCES

Butler, R. 1960. Intensive psychotherapy for the hospitalized aged. *Geriatrics* 15:644–653.

—— 1961. Re-awakening interests. *J. Amer. Nursing Home Assn.* 10:8–19.

Perlin, S. and R. Butler, 1964. Psychiatric aspects of adaptation to the aging experience. In National Institute of Mental Health, *Human Aging: Biological and Behavioral Aspects*. Washington, D.C.: U.S. Government Printing Office.

Rechtschaffen, A. 1959. Psychotherapy with geriatric patients: a review of the literature. *Journal of Gerontology* 14:73–84.

Late Adulthood
Relations with Others

Twenty-Nine

Interaction and Adaptation: Intimacy as a Critical Variable

Marjorie Fiske Lowenthal and Clayton Haven

In this selection, Lowenthal and Haven examine the importance of close friendship during late adulthood. Their findings suggest that the presence of a *confidant*—a close, intimate friend—may buffer the older adult against some of the adverse psychological consequences of certain stressful life events. In particular, adults who have the social support of a confidant appear to be better able to maintain high morale in the face of retirement and the loss of one's spouse than are individuals without such a relationship.

THIS PAPER is a sequel to previous studies in which we noted certain anomalies in the relation between traditional measures of social deprivation, on the one hand, and indicators of morale and psychiatric condition, on the other, in studies of older populations. For example, lifelong isolates tend to have average or better morale and to be no more prone to hospitalization for mental illness in old age than anyone else, but those who have tried and failed to establish social relationships appear particularly vulnerable (Lowenthal 1964). Nor, with certain exceptions, do age-linked trauma involving social deprivation, such as widowhood and retirement, precipitate mental illness (Lowenthal 1965). While these events do tend to be associated with low morale, they are by no means universally so. Furthermore, a voluntary reduction in social activity, that is, one which is not accounted for by widowhood, retirement, or physical impairment, does not necessarily have a deleterious effect on either morale or professionally appraised mental health status (Lowenthal and Boler 1965).

359

In analyzing detailed life histories of a small group of the subjects making up the samples for these studies, we were struck by the
fact that the happiest and healthiest among them often seemed to be
people who were, or had been, involved in one or more close personal
relationships. It therefore appeared that the existence of such a relationship might serve as a buffer against age-linked social losses
and thus explain some of these seeming anomalies. The purpose of
the present study is to explore this possibility. . . .

THE SAMPLE

The sample on which this report is based consists of 280 sample-
survivors in a panel study of community-resident aged, interviewed
three times at approximately one-year intervals. The parent sample
included 600 persons aged sixty and older, drawn on a stratified-
random basis from eighteen census tracts in San Francisco. The sample of 280 remaining at the third round of interviewing is divided about
equally in terms of the original stratifying variables of sex, three age
levels, and social living arrangements (alone or with others). As might
be expected, the sample differs from elderly San Franciscans (and
elderly Americans), as a whole, by including proportionally more of
the very elderly, more males, and more persons living alone. Largely
because of the oversampling of persons living alone, the proportion
of single, widowed, and divorced persons, and of working women is
higher than among elderly Americans in general. Partly because of
the higher proportion of working women the income level of the sample
was higher than average (44 percent having an income of over $2,000
per year, compared with 25 percent among all older Americans). The
proportion of foreign born (34 percent) resembles that for older San
Franciscans in general (36 percent), which is considerably more than
for all older Americans (18 percent). While some of these sample
biases may tend to underplay the frequency of the presence of a
confidant, we have no reason to believe that they would influence
findings from our major research question, namely, the role of the
confidant as an intervening variable between social resources and
deprivation and adaptation.

Measures of Role, Interaction and Adaptation

The two conventional social measures to be reported here are number of social roles[1] and level of social interaction.[2] Men tend to rank somewhat lower than women in social interaction in the younger groups, but up through age seventy-four, fewer than 17 percent of either sex rank "low" on this measure (defined as being visited [only] by relatives, contacts only with persons in dwelling, or contacts for essentials only). Isolation increases sharply beginning at age seventy-five, however, and in that phase women are slightly more isolated than men (32 percent as compared with 28 percent), possibly because of the higher proportion of widows than of widowers. In general, because they are less likely to be widowed and more likely to be working, men have more roles than women at all age levels, and this discrepancy becomes particularly wide at seventy-five or older, when 40 percent of the men, compared with only 15 percent of the women, have three or more roles. Among men, a low level of social interaction and a paucity of social roles tend to be related to low socioeconomic status, but this does not hold true for women.

The principal measure of adaptation in this analysis is a satisfaction-depression (or morale) score based on a cluster analysis of answers to eight questions.[3] For a subsample of 112, we shall also report ratings of psychiatric impairment by three psychiatrists, who, working independently, reviewed the protocols in detail but did not see the subjects.[4] We thus have a subjective indicator of the sense of well-being and a professional appraisal of mental health status. A third measure, opinion as to whether one is young or old for one's age, is included to round out the adaptation dimension with an indication of what might be called the respondent's opinion as to his relative deprivation—that is, whether he thinks he is better or worse off than his age peers. . . .

[1] Roles include parent, spouse, worker, church-goer, and organization member.
[2] Ranging from "contributes to goals of organizations" to "contacts for the material essentials of life only." All measures of interaction and adaptation reported here pertain to the second round of follow-up.
[3] The distribution of individual cluster scores was dichotomized at the median: persons falling below the median are called "depressed," and persons falling above the median are called "satisfied." Questions pertained to the sense of satisfaction with life, happiness, usefulness, mood, and planning.
[4] One-third (38 persons) were judged impaired; the majority (30 persons) only mildly so. Seven were rated moderately and one severely impaired.

Pursuing our hypothesis with regard to the potential importance of a confidant as a buffer against social losses, we turn to table 29.1, which shows the current presence or absence of a confidant, and recent losses, gains, or stability in intimate relationships, in conjunction with the three adaptation measures. As the first column shows, the presence of a confidant is positively associated with all three indicators of adjustment. The absence of a confidant is related to low morale. Lack of a present confidant does not, however, have much bearing on the individual's sense of relative deprivation or on the psychiatric judgments of mental impairment. We suggest, though with the present data we cannot fully document, two possible explanations for these findings. First, as we have previously noted (Lowenthal 1964), there are some lifelong isolates and near-isolates whose late-life adaptation apparently is not related to social resources. The sense of relative deprivation, at least for older persons, no doubt applies not only to current comparisons with one's peers, but also to comparisons with one's own earlier self. This would contribute to an explanation of the fact that some older people without a confidant are satisfied. They do not miss what they have never had. Our second explanation echoes the old adage that it is better to have loved and lost than never to have loved at all. The psychiatrists, in rating mental health status, may take *capacity* for intimacy into account, as indicated by past relationships such as marriage or parenthood. The respondent, in a more or less "objective" comparison of himself with

Table 29.1. The Effect of a Confidant on Adaptation

| | Current Presence of Confidant | | Change in Confidant Past Year | | |
	Yes %	No %	Gained %	Lost %	Maintained Same Confidant %
Psychiatric status					
Unimpaired	69	60	(60)	(56)	80
Impaired	31	40	(40)	(44)	20
Opinion of own age					
Young	60	62	42	49	61
Not young	40	38	58	51	39
Morale					
Satisfied	59	41	44	30	68
Depressed	41	59	56	70	32

his peers, may also take these past gratifications into account. However, such recollections may well be less serviceable on the more subjective level of morale and mood.

The right side of the table, showing change and stability in the confidant relationship, dramatically exemplifies the significance of intimacy for the subjective sense of well-being: The great majority of those who lost a confidant are depressed, and the great majority of those who maintained one are satisfied. Gaining one helps, but not much, suggesting again the importance of stability, which we have noted in relation to the other social measures. Maintenance of an intimate relationship is also strongly correlated with self-other comparisons and with psychiatrists' judgments, though losses do not show the obverse. This supports our suggestion that evidence of the *capacity* for intimacy may be relevant to these two more objective indicators of adaptation.

The great significance of the confidant from a subjective viewpoint, combined with the fact that sizable proportions of people who reduced their social interaction or social role nevertheless were satisfied, raised the possibility that the maintenance of an intimate relationship may serve as a buffer against the depression that might otherwise result from a lessened social role or interaction, or from the more drastic social losses frequently suffered by older persons, namely, widowhood and retirement. To test this hypothesis, we examined the morale of those who changed on the interaction and role measures in the light of whether they did or did not have a confidant.

The Intimate Relationship as a Buffer Against Social Losses

As table 29.2 shows, it is clear that if you have a confidant, you can decrease your social interaction and run no greater risk of becoming depressed than if you had increased it. Further, if you have no confidant, you may increase your social activities and yet be far more likely to be depressed than one who has a confidant but has lowered his interaction level.[5] Finally, if you have no confidant and retrench in your social life, the chances for depression become over-

---

[5] Parallel analyses of the other two adjustment measures are not included here. The psychiatric ratings are available for only a subsample of 112 (and cells would become too small). The indicator of opinion of own age reflects trends similar, though not so marked, as those shown here, except for increase in role status, where the absence of a confidant does not contribute to a negative opinion.

Table 29.2 Effect on Morale of Changes in Social Interaction and Role Status, in the Presence and Absence of a Confidant

	Increased		Decreased	
	Has Confidant %	No Confidant %	Has Confidant %	No Confidant %
	Social Interaction			
Morale				
Satisfied	55	(30)	56	13
Depressed	45	(70)	44	87
	Role Status			
Morale				
Satisfied	55	(42)	56	(38)
Depressed	45	(58)	44	(62)

whelming. The findings are similar, though not so dramatic, in regard to change in social role: if you have a confidant, your roles can be decreased with no effect on morale; if you do not have a confidant, you are likely to be depressed whether your roles are increased or decreased (though slightly more so if they are decreased). In other words, the presence of an intimate relationship apparently serves as a buffer against such decrements as loss of role or reduction of social interaction.

What about the more dramatic "insults" of aging, such as widowhood or retirement? While a few people became widowed or retired during our second follow-up year (and are therefore included among the "decreasers" in social role), there were not enough of them to explore fully the impact of these age-linked stresses. Therefore, we checked back in the life histories of our subjects and located those who retired within a seven-year period prior to the second follow-up interview or who became widowed within this period. Though our concern is primarily with social deficits, we added persons who had suffered serious physical illness within two years before the second follow-up contact since we know that such stresses also influence adaptation.

Table 29.3 indicates that the hypothesis is confirmed in regard to the more traumatic social deprivations. An individual who has been widowed within seven years, and who has a confidant, has an even higher morale than a person who remains married but lacks a con-

Table 29.3. Effect of Confidant on Morale in the Contexts of
Widowhood, Retirement, and Physical Illness

	Satisfied	Depressed
Widowed within 7 years		
Has confidant	55	45
No confidant	(27)	(73)
Married		
Has confidant	65	35
No confidant	(47)	(53)
Retired within 7 years		
Has confidant	50	50
No confidant	(36)	(64)
Not retired		
Has confidant	70	30
No confidant	50	50
Serious physical illness within 2 years		
Has confidant	(16)	(84)
No confidant	(13)	(87)
No serious illness		
Has confidant	64	36
No confidant	42	58

fidant. In fact, given a confidant, widowhood within such a comparatively long period makes a rather undramatic impact on morale. Among those having confidants, only 10 percent more of the widowed than of the married are depressed, but nearly three-fourths of the widowed who have no confidant are depressed, compared with only about half among the married who have no confidant.[6] The story is similar with respect to retirement. The retired with a confidant rank the same in regard to morale as those still working who have no confidant; those both retired and having no confidant are almost twice as likely to be depressed as to be satisfied, whereas among those both working and having a confidant, the ratio is more than reversed.

Although relatively few people (35) developed serious physical illness in the two-year period prior to the second follow-up interview, it is nevertheless amply clear that a confidant does not play a mediating role between this "insult" of aging and adjustment as measured by the depression-satisfaction score. The aged person who is (or has recently been) seriously ill is overwhelmingly depressed, re-

[6] This finding suggests the need for far more detailed questioning on the confidant relationship than we were able to undertake. It may well be that some married persons assumed that the question pertained to confidants other than spouses.

gardless of whether or not he has an intimate relationship. Superficially, one might conclude that this is a logical state of affairs. Social support—such as an intimate relationship—may serve as a mediating, palliative, or alleviating factor in the face of social losses, but one should not expect it to cross system boundaries and serve a similar role in the face of physical losses. On the other hand, why doesn't one feel more cheerful, though ill, if one has an intimate on whom to rely for support or to whom one can pour out complaints? At this point we can only conjecture, but one possible explanation is that serious physical illness is usually accompanied by an increase in dependence on others, which in turn may set off a conflict in the ill person more disruptive to his intimate relationships than to more casual ones. This may be especially true of dependent persons whose dependency is masked (Goldfarb 1965). A second possibility is that the assumption of the sick role may be a response to the failure to fulfill certain developmental tasks. In this event, illness would be vitally necessary as an ego defense, and efforts of intimates directed toward recovery would be resisted (Lowenthal and Berkman, 1967). A third possibility is that illness is accompanied by increased apprehension of death. Even in an intimate relationship, it may be easier (and more acceptable) to talk about the grief associated with widowhood or the anxieties or losses associated with retirement than to confess one's fears about the increasing imminence of death. . . .

IMPLICATIONS

The maintenance of a stable intimate relationship is more closely associated with good mental health and high morale than is high social interaction or role status, or stability in interaction and role. Similarly, the loss of a confidant has a more deleterious effect on morale, though not on mental health status, than does a reduction in either of the other two social measures. We suggested that while psychiatrists may take the capacity for an intimate relationship into account in their professional judgments, awareness that he has such a potential does not elevate an individual's mood if he has recently lost a confidant. Finally, we have noted that the impact on adjustment of a decrease in social interaction, or a loss of social roles, is considerably softened if the individual has a close personal relationship. In addition, the age-linked losses of widowhood and retirement are also ameliorated by the presence of a confidant, though the assault of physical illness clearly is not.

REFERENCES

Goldfarb, A. 1965. Psychodynamics and the three-generation family. In. E. Shanas and G. Streib, eds., *Social Structure and the Family: Generational Relations*, pp. 10–45. Englewood-Cliffs, N.J.: Prentice-Hall.

Lowenthal, M. F. 1964. Social isolation and mental illness in old age. *American Sociological Review* 29:54–70.

—— 1965. Antecedents of isolation and mental illness in old age. *Archives of General Psychiatry* 12:245–254.

Lowenthal, M. F. P., Berkman et al. 1967. *Aging and Mental Disorder in San Francisco: A Social Psychiatric Study*. San Francisco: Jossey-Bass.

Lowenthal, M. F. and D. Boler. 1965. Voluntary vs. involuntary social withdrawal. *Journal of Gerontology* 20:363–371.

Thirty

The Meaning of Friendship in Widowhood
Helena Znaniecki Lopata

Friendship during the late adulthood years may be especially important for women since, by and large, they are more likely than men to outlive their spouse. In this selection, Helena Lopata discusses the meaning and significance of friendship among widows. Her findings show that a surprising number of widows are socially isolated. This isolation tends to be especially true among the least educated and most economically disadvantaged.

AMERICANS ARE CONFUSED about friendship. We have idealized it, yet we view it with caution, even with fear. We conjure images of fraternal (more often than sororal) bonds, of "buddy" comradeship, of the sharing of intimate secrets, and of perfect symmetry of commitment between people equal in power and status (Little 1970; Lowenthal and Haven 1968). This equality is supposed to be reinforced by similarity of background and a sharing of values. Friends are supposed to be willing to sacrifice everything, even their lives, for each other and to put their friendship ahead of all other relationships. At the same time, we don't think our own lives can contain such a deep relationship because there are other roles we must perform. We may feel that we once had such a friendship, maybe even more than one, when we were young—and we may hope that we can have it again in the future, when we have time for it.

We suspect that friendship, in its ideal form, could lead to disloyalty to family or employer. Although we spend millions of dollars in learning "how to win friends and influence people" (Carnegie 1936), we worry about the consequences of friendship among adults. Cross-sex intimacy is presumed to lead inevitably to sexual intercourse. Same-sex friendships for either man or woman are seen as potentially interfering with the marital bond. Since husbands and wives are considered perfect companions, neither should need another close friend.

368

Women must "hang loose" in their relationships so as not to interfere with their husband's career and their shared social life. A wife must be ready to move, leaving friends behind, when her husband's job calls for a transfer or when his success calls for a new life-style and new associates with which to share it. Old friends are at a distance, not just geographically but socially. Husbands must be ready to move up to the next rung on occupational ladders and must not develop close ties at work that need to be broken (Whyte 1956). They must also be careful whom they associate with away from the job. Above all, it is dangerous to exchange confidences—people may use these against you. Wives also must watch what they say and to whom they say it so as not to endanger their husband's bread-winning responsibility. . . .

In one group in our society friendship has been fully encouraged: in older people. They are no longer supposed to be involved in the basic roles of breadwinner and spouse of breadwinner and are freed from worry about the dangers friendship could impose on their lives. The older woman is free from the obligation to have and care for children and, if her husband dies, even from her role of wife. Her husband, if he survives till retirement, need no longer worry about developing disadvantageous alliances or about disclosure of information that could hurt his job or career. Old men and women are supposed to be less interested in sex, so even cross-sex friendships are permissible.

Of course, the encouragement of friendship among older Americans may not necessarily lead to their formation, at least in their idealized form (Blau 1961, 1973). In order to determine the actual importance of friendship in the lives of older Americans, we can examine the way it fits into the support systems of urban widows. A support system includes the exchange of an action or an object that the giver and/or receiver defines as necessary or helpful in maintaining a style of life. Each person is involved in various forms of support exchanges—financial, service, social, or emotional (Lopata 1975a).

FRIENDSHIP IN WIDOWHOOD

In order to learn the importance of friendship in the lives of women who no longer have to be devoted to either the role of wife or mother, we studied over 1,000 widows living in the Chicago area. These women represented over 82,000 widows who were or had recently been beneficiaries of Social Security. We can predict two opposite possibilities. On the one hand, these widows might become heavily oriented toward friendship, both deepening relations with old friends and developing new friends. They might engage in a variety of social activities, turning to others for emotional and even service supports. On the other hand, these women might be more apt to follow the personality and life-style habits of their past, so that those who had relied little on friends in the past would not become friend-oriented when their roles of wife and mother decreased in importance. People tend to develop a style of dealing with the world that continues throughout life, albeit with modifications introduced by changing definitions or circumstances. Thus, in spite of the fact that widows are free to make their own friendships, it is quite possible that they will follow the pattern they developed while their husbands were still living. What they do with their friends, how much time they spend with them, and even those persons they are friendly with might change with widowhood, but women who have been strongly involved in friendships will probably continue to do so and the friendless will continue being friendless. Conditions such as education, income, and race can be expected to affect friendship now as they did when the women were married (Lopata 1973a, 1973b, 1975b). We can expect widows to be friends mostly with other widows (Lopata 1973a; Rosow 1967; Blau 1973; Lowenthal 1968). Strains could develop in relations with people who are still married. Widows have more in common with other widows, and they have little opportunity to continue the style of social life that is based on couples (Cumming and Henry 1961). Finally, we can expect that close personal friends will appear as service, social, and emotional supports of widows.

The life-pattern expectations of friendships are supported by our widowhood study. Very few of those widows who said they had no friends the year before their husband's fatal illness or accident established new friendships after he died. Over one-third of the women kept their old friends but did not establish new ones. Another third said they have both old and new friends. This leaves a surprising one-sixth of the widows who claimed to be totally without friends both when their husband was still well and at the time of the study.

THE FRIENDLESS WIDOW

The completely friendless women were most apt to have been born outside Chicago, migrating from another country or from American rural areas. An even larger proportion had nonurban parents. Proportionately more black than white widows claimed to have no close friends, and there was a strong association between the number of years of formal education and whether they had friends and how many. The amount of formal schooling affected whom they married, their husband's background, and the jobs he had during his lifetime. The job, in turn, affected family income and life-style. Furthermore, the older the women, the more likely they were to have suffered a double burden. They were likely to have been immersed in a culture suspicious of nonrelative "strangers," one that viewed the world as an unfriendly place. They were also likely to have experienced a lifestyle that provided little time, energy, or money for leisure-time pursuits or had the advantages of abundance that make pleasurable friendship possible (Znaniecki 1965, Lopata 1969, 1971, 1973a, 1973b). They are now living in a situation in which the traditional support systems of family and neighborhood have dissolved, yet they are unaccustomed to going out to look for and develop new social relations and social roles. They either never develop social networks outside the family, or they cannot reengage in society once former networks are disorganized by death, the dispersal of siblings or children, or financial or health changes. These are the more isolated women. Their lives were peripheral to the modern urban social world even when their husbands were living. They are even more isolated after their husbands' death.

The friendless women do not necessarily say they want friends. They are likely to believe that "relatives are your only true friends" (Lopata 1973a, 1975b). Over half of these widows in the recent study who claimed they had no personal friends when their husbands were alive and none since their death disagree with the statement "I wish I had more friends." Nor does the absence of friends automatically lead to dissatisfaction with life or loneliness. When asked to define their level of loneliness in comparison to what they assume is true of other people, over half place themselves as "less lonely than most people," "rarely" or "never" lonely. Since loneliness is often a matter of relative deprivation, in that the person experiencing it compares her or his level of social interaction with a higher level experienced in the past or assumed to be experienced by others, we are led to believe that these women do not expect social interaction with people identified as friends (Lopata 1969, 1973b; Weiss 1973).

There seem to be two types of friendless widows (probably friendless people in general). One type does not identify friendship as part of life, does not expect to have friends, and is not lonely or dissatisfied with life in their absence. The other type wants friends and is lonely and dissatisfied with life. This second type is apt to lack the self-confidence, the "aggressive" initiative, or the knowledge of how to venture out of her environment, even after it becomes disorganized by the death of her husband. Such a widow is likely to sit at home passively, waiting for someone else to initiate contacts and interaction. Unfortunately, there are few people who have the time or inclination to go through the laborious process of converting such a woman into a socially involved person. Other lonely widows, who could become friends because of shared interests, also lack the resources for doing so. There are undoubtedly many nonwidowed women who have the same lack of personal resources for building social networks.

THE WOMAN WITH PRE-WIDOWHOOD FRIENDS

By contrast, women who stated that they had close peronsal friends the year before their husbands' death, and especially those who listed more than one such intimate, were found to be younger, both at the time of the study and the time of widowhood, to be better educated, with higher incomes when their husbands were living and also at the time of the study, and to be white. They themselves were apt to have been working just before or during widowhood and to have held white-collar jobs. Their late husbands were also likely to have been in white-collar occupations.

Three-fourths of all the widows listed at least one friend before their husbands' fatal illness or accident, with an average of almost three per woman. Most of these friends were women, but men were sometimes mentioned, usually as part of a couple. Couple friendship is characteristic of the American middle class, whose social life is, in fact, embedded in couple-companionate, leisure-time interaction (Znaniecki 1965; Lopata 1971, 1973a, 1975b). Widows seldom listed a man without his wife, again reflecting the traditional discouragement of individualized cross-sex friendship.

Most of the pre-widowhood friendships mentioned were long-standing, known for more than ten years before the husband's death.

Almost all of the people were married at that time, which supports the couple-companionate thesis (Hunt 1966; Blau 1973). Contrary to the findings of earlier studies of friendship among couples (Babchuk 1965; Babchuk and Bates 1963), in which the husband was said to have selected the couples with which they were friends, the Chicago-area widows took the credit for developing their couple friendships. They said they did so by taking advantage of contacts in their neighborhood, through voluntary associations, or at work. Sometimes they found friends through relatives or even in such esoteric places as bars or doctors' offices. Younger women were still likely to have friends from childhood.

Most of the pre-widowhood friends were seen an average of once a week, though there were racial and educational differences. The more socially involved women, who had more friends, often had a lower average frequency of contact with their friends when their husbands were still alive. However, few of the widows were able to maintain the same frequency of contact with their old friends. At least, by the time they were being interviewed, the frequency had dropped to "several times a year," although most old friends were still considered friends. Changed circumstances of life often make contact difficult, but a major problem, particularly among the middle-class women accustomed to couple-companionate interaction, is the asymmetry produced by a husband's death. Widows report feeling like a "fifth wheel" when they try to go out with or go to the home of couple friends unless they "have a date" (Lopata 1973a, 1975a). To make matters worse, new widows are not only husbandless and escortless, but they are apt to be grieving, which makes friends feel awkward and uncomfortable. Their friends complain that they do not respond to friendly advances, the relations become strained, the widows' lowered self-confidence makes them oversensitive, and they drift apart. Whoever is at fault, contacts with married friends were found to decrease, particularly when the husbands were present.

Friends who were not tied to a couple-companionate round of activities were found to be more easily retained in widowhood. Of course, some of these old friends had themselves become widowed, in which case contact often increased. The widows who reported themselves as the "most lonely person" they know or "more lonely than most people" had experienced a dramatic drop in contacts with old friends and an inability to replace them with new friends. Thus, the most lonely were the women who lost not just a husband but also friends from the past, both in terms of closeness and in frequency of interactions.

THE WOMAN WITH POST-WIDOWHOOD FRIENDS

Over half of the widows in the study claimed they had not made new friends since their husbands died. Again, there was a strong association between education and the making of new friends. The more educated developed more friends. Current total household income proved to be a greater influence on the ability to develop new friendships than did prior income, partly because many widows experienced a considerable drop of income with the death of their husbands. It was reported that new friends were also met primarily in the neighborhood, through voluntary associations, or on the job—the same sources of contacts as before widowhood. Most of these new friends were married, mainly because most women in the younger age groups are still in that marital status, but there were many more who were widowed. There was also an increase in men listed without a wife, and some men were even identified as "boyfriends." A few of the widows in the study have already remarried.

CONTRIBUTIONS OF FRIENDS TO THE SUPPORT SYSTEMS OF WIDOWS

Generally speaking, those women who listed old friends or developed new friends found them helpful during the time when they were rebuilding their lives following the death of their husbands and the heavy grief period. Only one-tenth of the widows with friends evaluated them as rarely or never helpful. In fact, friends were judged as helpful during this traumatic time of life more often than any other significant associates except adult children.

In view of the frequency with which these widows listed friends, particularly old friends, as very helpful during the time they were rebuilding their lives, it is surprising to find that friends did not appear as frequent contributors to their support systems. We listed about 65 different economic, service, emotional, and social supports, giving each woman the opportunity to name three persons for each support. This means that the widow could have listed as many as 195 names, although she could give the same person in several different supports. The list included emotional supports the year before the husband's death as well as at the time of the study. Economic and service supports being examined included both an inflow, with the widow as the recipient, and an outflow, with her as the giver. The social supports asked about specified social activities that people often share with others.

Friends were almost totally absent from the list of people providing economic supports, as givers or receivers of money or gifts; as helpers in payment of rent, food, or clothing; or as helpers in meeting other bills. In fact, the widows were relatively independent financially. Friends were also relatively absent from service supports such as helping with household repairs, shopping, housekeeping, yard work, child care, car care, decision-making, legal aid, or transportation.

Friends did appear, however, in the social support system. We can conclude from this that if widows engage at all in typical urban American social activities, they often do so with people listed as friends. Although half of our population does not go to public places like movie theaters, over one-third of those who do, go with a friend. Again, although one-fifth of the widows claimed never to visit anyone, four in ten of those visited were friends. They were the people whom the widows were most likely to entertain or share lunches with. Playing cards or engaging in sports or other games is also a friend-sharing activity; over half of the companions listed for these activities were friends. It must be noted, however, that over half of the women in the study had never engaged in such activities. On the other hand, friends were found to be much less likely to be companions in church attendance, in travel out-of-town, or in celebrating holidays. All in all, many women appeared to be remarkably restricted in their social activities. Polite companionship (Znaniecki 1965) and leisure-time interaction (Dulles 1965) do not seem to describe the social life of many widows, particularly the older ones with less education and more of a blue-collar life-style.

In spite of the relative prominence of friends in the social support system of socially active widows, they were largely absent in the emotional support system. Less than one in ten of the people named as those the women felt closest to the year before their husbands' death were friends. Almost one in four named her husband as the person she had felt closest to, and one-third of the women named their children. Moreover, their husbands' death did not push many women toward friends. Only one-tenth of the "closest" references during the time of the study were listed as friends. The same distribution occurred when we asked the widow whom she most enjoyed being with when her husband was well and whom she most enjoyed being with at the time the study was conducted. A friend was not even the main confidante for most widows. About one-tenth said that they told their problems to a friend either before widowhood or now. Even fewer women said they turned to their friends for comfort when they were feeling blue or for help in times of crises. Few got the feeling of being an important person from their friends. At the same time, few said friends were the people they were most often angry at. When the researchers asked, "Now I am going to read some 'feeling states' which many people think are important for a full life. What persons or groups made (make) you feel this way in 19___ (the year before the late husband's death)/now?" Few women mentioned friends. The feeling of being "respected" drew less than one-tenth of the references to friends now and an even smaller proportion for the "before" period; the feelings of being "useful," "independent," "self-sufficient," or "secure" drew very few friends for the "before" and just slightly more for the "now" period. Only the feeling of being "accepted" drew more than one-tenth of the references. Most women who turned to friends to meet emotional needs did so only after having listed someone else, usually their husbands when they were alive and their children now. Children rather than friends are looked to for most emotional support.

SUMMARY AND CONCLUSIONS

However idealized friendship is in American society and however women are assumed to turn to friendship when they are freed from the obligations and involvements of their basic roles of wife and mother, most Chicago-area widows were not found to follow this pattern. Although some women in the study were deeply involved in their relations with friends, the surprising finding of the widowhood study was that so few were. Some widows appeared to be socially isolated. They had no friends before their husbands' death and they have no friends now that they are widows. Other women had a limited number of friends, almost invariably other women, when their husbands were living and they were able to bring these sex-segregated friendships into widowhood. There were widows who reflected the style of life described by Whyte (1956), in which most social relations are built around the man's job and career. They tended to associate almost exclusively with other couples and engaged in couple-companionate interaction, usually during evenings and especially weekends. These women had difficulties maintaining such friendships into widowhood. Being the "fifth wheel" on social occasions, often feeling an unwillingness on the part of married women friends to include them, they were left out of that type of interaction. Then they found that they had to convert their social life into sharing lunches or other activities that women engage in together and remain alone during the times that call for couple sharing. Other widows, usually the younger ones, obtained escorts or even new husbands and rejoined old groups or formed a new couple-companionate circle.

The social superficiality of relations considered "close friendships" emerges when we try to determine if people listed as friends are actually involved in the support systems of wives and later, widows. It is here we see that undertaking social activities with old or new friends does not convert itself into deeper involvement in the economic, service, or even emotional support systems. When given a chance to list as many as three persons closest to her, to whom she tells problems, and who comfort her when she is blue, nine-tenths of the Chicago-area widows did not even think of the friends they had listed as "close personal friends." Apparently, Americans' cautious attitudes about friends prevent even widows from forming close ties with them.

However, the distribution of friendships within our population of widows leads to some predictions as to this relationship in the future. Friendships were fewer and more superficial among the least educated and the most disadvantaged of the Chicago-area women. More white-collar women, even those near the border of poverty in widowhood, wished for friendships and had the personal resources for developing them. Thus, the blue-collar or working-class ideology that identifies relatives as the only possible friends for women, together with social sanctions against converting strangers into social companions, seems to restrict the friendships of women more than does the middle-class ideology that cautions women to avoid strong friendships because of their husbands' career. Should present middle-class values predominate, we could expect to find more true friendships among women, married or not. This does not solve the problems of the widow, accustomed to couple interaction and now deprived of an escort needed to participate in it. Nor does it guarantee friendship involvement beyond social support systems. If the feminist movement increases its influence down through the American class system, we can expect that women will receive greater satisfaction from friendship with other women, and that widows will not feel a stigma because they are husbandless. Feminism may also lead to greater depth in woman-to-woman relationships.

REFERENCES

Babchuk, Nicholas. 1965. Primary friends and kin: a study of the associations of middle-class couples. *Social Forces* 43:483–493.

Babchuk, Nicholas with A. P. Bates. 1963. Primary relations of middle-class couples: a study of male dominance. *American Sociological Review* 28:374–384.

Blau, Zena. 1961. Structural constraints of friendship in old age. *American Sociological Review* 26:429–439.

—— 1973. *Old Age in a Changing Society*. New York: Franklin Watts.

Carnegie, Dale. 1936. *How to Win Friends and Influence People*. New York: Simon and Schuster.

Cumming, Elaine and William E. Henry. 1961. *Growing Old: The Process of Disengagement*. New York: Basic Books.

Dulles, Foster Rhea. 1965. *A History of Recreation*. New York: Appleton Century-Crofts.

Hunt, Morton M. 1966. *The World of the Formerly Married*. New York: McGraw-Hill.

Little, Roger. 1970. Buddy relations and combat performance. In Oscar Grusky and George A. Miller, eds., *The Sociology of Organizations: Basic Studies*. New York: Free Press.

Lopata, Helena Z. 1969. Loneliness: forms and components. *Social Problems* 17:248–262. Reprinted in Robert S. Weiss, ed., *Loneliness: The Experience of Emotional and Social Isolation*. Cambridge, Mass.: MIT Press, 1973.

—— 1971. *Occupation: Housewife*. New York: Oxford University Press.
—— 1973a. *Widowhood in an American City*. Cambridge, Mass.: Schenkman Publishing Co., General Learning Press.
—— 1973b. The effect of schooling on social contacts of urban women. *American Journal of Sociology* 79:604–619.
—— 1975a. Support systems of widows. Report to the Social Security Administration.
—— 1975b. Couple companionate relationships in marriage and widowhood. In Nona Glazer Malbin, ed., *Old Family/New Family*. New York: Van Nostrand.
Lowenthal, Marjorie and C. Haven. 1968. Interaction and adaptation: intimacy as a critical variable. *American Sociological Review* 33:20–30.
Packard, Vance. 1962. *The Pyramid Climbers*. Greenwich, Conn.: Fawcett Crest.
Rosow, Irving. 1967. *The Social Integration of the Aged*. New York: Free Press.
Weiss, Robert S. 1973. *Loneliness: The Experience of Emotion and Social Isolation*. Cambridge, Mass.: MIT Press.
Whyte, William H., Jr. 1956. *The Organization Man*. New York: Simon and Schuster.
Znaniecki, Florian. 1965. *Social Relations and Social Roles*. San Francisco: Chandler.